THE FOUNDATIONS OF DEFENSIVE DEFENCE

W0106550

Also by Anders Boserup

CHEMICAL AND BIOLOGICAL WARFARE AND THE LAW OF WAR
WAR WITHOUT WEAPONS: Non-Violence in National Defence
  (*with Andrew Mack*)
IMPLICATIONS OF ANTI-BALLISTIC MISSILE SYSTEMS (*co-editor*)
THE CHALLENGE OF NUCLEAR WEAPONS (*co-editor*)

Also by Robert Neild

PRICING AND EMPLOYMENT IN THE TRADE CYCLE
THE MEASUREMENT AND REFORM OF BUDGETARY POLICY
  (*with T. S. Ward*)
HOW TO MAKE UP YOUR MIND ABOUT THE BOMB
* AN ESSAY ON STRATEGY

* *Also published by Macmillan*

The Foundations of Defensive Defence

Edited by

Anders Boserup
Director, European Centre for International Security
Copenhagen

and

Robert Neild
Professor Emeritus of Economics, University of Cambridge
and Fellow of Trinity College, Cambridge

Assisted by David Carlton

MACMILLAN

First published 1990

Published by
THE MACMILLAN PRESS LTD
Houndmills, Basingstoke, Hampshire RG21 2XS
and London
Companies and representatives
throughout the world

Typeset by
Footnote Graphics, Warminster, Wiltshire

British Library Cataloguing in Publication Data
The foundations of defensive defence.
1. Europe. Western Europe. Defence. Policies of
governments
I. Boserup, Anders *1940–*   II. Neild, R. R. (Robert Ralph) *1924–*
III. Carlton, David, *1938–*
355'.0094
ISBN 978-0-333-52999-7   ISBN 978-1-349-20733-6 (eBook)
DOI 10.1007/978-1-349-20733-6

Contents

Preface

The origin of this collection of writings on the theory of defensive force structures and its application to conventional disarmament in Europe is a series of meetings of a Workshop on Conventional Forces in Europe organised by the Pugwash Conferences on Science and World Affairs. These Pugwash Workshops, of which seven have so far been held since 1984, can be said to have played a pioneering role in generating a debate about defensive strategy between scholars and military professionals from East and West and in formulating ideas which are now coming on to the agenda of policy-makers.

This is not a reproduction of the full proceedings of the meetings of the workshop. In order to provide a comprehensive but economical view of the subject, the volume contains a selection of the papers circulated to the workshop plus a few papers from other sources. An account of the background to the workshop and of the ideas it set out to explore is to be found in the Introduction.

We are most grateful to the Joseph Rowntree Charitable Trust for providing a grant to cover the expenses of editing a linguistically heterogeneous collection of papers, and to David Carlton who undertook that task. For typing we are grateful to Linda Freeman.

ANDERS BOSERUP
ROBERT NEILD

List of Abbreviations

APC	Armoured Personnel Carrier
ATGW	Anti-tank Guided Weapon
ATTU	Atlantic to the Urals
CDE	Conference on Disarmament in Europe
CEPs	Circular Errors Probable
CFE	Conventional Forces in Europe (Talks)
CSBM	Confidence and Security-building Measure
CSCE	Conference on Security and Cooperation in Europe
CST	Conventional Stability Talks
C^3I	Command, Control, Communications and Intelligence
DEH	Defence Efficiency Hypothesis
FDF	Force Ratio Development Function
FOFA	Follow-on Forces Attack
GDR	German Democratic Republic
INF	Intermediate-range Nuclear Forces
MBFR	Mutual Balanced Force Reduction (Talks)
MBT	Main Battle Tank
MFR	Mutual Force Reduction (Talks)
MLRS	Multiple-launcher Rocket System
NATO	North Atlantic Treaty Organisation
PGM	Precision-guided Munition
R and D	Research and Development
SACEUR	Supreme Allied Commander Europe (NATO)
SALT	Strategic Arms Limitation Talks
SDI	Strategic Defence Initiative
SLBM	Submarine-launched Ballistic Missile
SLCM	Surface-to-surface Missile
START	Strategic Arms Reduction Talks
TNF	Theatre Nuclear Forces
VSTOL	Vertical or Short Take-off/and Landing

Notes on the Contributors

Alexei Arbatov (*Soviet*) is a Head of Department in the Institute for World Economic and International Affairs which operates under the auspices of the Academy of Sciences.

Georgy Arbatov (*Soviet*) is Director of the Institute of United States and Canadian Studies in Moscow and a full member of the Academy of Sciences. Since 1981 he has been a Member of the Central Committee of the Communist Party of the Soviet Union and was elected as a People's Deputy in March 1989.

Sir Hugh Beach (*British*) is a general who, after retirement, served as Director of the Council for Arms Control in London from 1986 to 1989.

Anders Boserup (*Danish*) is Director of the Copenhagen section of the European Centre for International Security and co-convenor of the Pugwash Study Group on Conventional Forces in Europe.

Jonathan Dean (*US*) is Arms Control Adviser to the Union of Concerned Scientists. From 1973 to 1981 Deputy Head, then Head, of the US Delegation to the negotiations on Mutual and Balanced Force Reductions.

Jens Joern Graabaek (*Danish*) is a major in the Royal Danish Air Force, who since 1985 has belonged to the independent Advisory and Analysis Group of the Danish Minister of Defence.

Hans Hofmann (*West German*) is a Professor for Operations Research in the Computer Science Department of the Federal Armed Forces University in Munich since 1982.

Reiner Huber (*West German*) is Professor of Applied Systems Science at the Federal Armed Forces University in Munich.

Carlo Jean (*Italian*), a serving general in the Italian Army, is Director of the Military Centre for Strategic Studies in Rome and a lecturer at the Alcide de Gasperi Graduate School in Rome.

Andrzej Karkoszka (*Polish*) is a Research Fellow at the Polish Institute of International Affairs in Warsaw since 1969.

Andrey Kokoshin (*Soviet*) is a Corresponding Member of the USSR Academy of Sciences and is Deputy Director of the Institute of United States and Canadian Studies, Moscow.

Alexander Konovalov (*Soviet*) is Head of Section in the Department of Military–Political Studies of the Institute of United States and Canadian Studies, Moscow.

Valentin Larionov (*Soviet*), a retired General, is a consultant to the Institute of United States and Canadian Studies, Moscow. He is a well-known military writer and a former Professor at the General Staff Academy, Moscow.

Albrecht von Müller (*West German*) is director of the European Centre for International Security, Starnberg, and co-convenor of the Pugwash Study Group on Conventional Forces in Europe.

Robert Neild (*British*) is Professor Emeritus of Economics at the University of Cambridge and a Fellow of Trinity College, Cambridge.

Elmar Schmähling (*West German*), is a rear-admiral, and is head of the Office for Studies and Exercises of the West German Federal Armed Forces.

1 Introduction
Anders Boserup and Robert Neild

The chapters in this volume explore and expound a strategic idea that has been important in the easing of the military confrontation between NATO and the Warsaw Pact. It is an idea which had its roots in the West, was taken up in the East as part of the new thinking promoted by Mikhail Gorbachev, and is now, in different forms, part of the dialogue between the two sides.

The idea is to ease the military confrontation in Europe by restructuring conventional forces so as to minimise their capability to attack while maintaining intact their capability to defend. If that can be done, it will provide unambiguous evidence of peaceful intentions; it will be mutually reassuring; and it will enhance military stability. This sounds a simple and easily acceptable idea, but it is neither: it is not simple to put into practice; and since it represents a reversal of previous thinking, it is not easily accepted.

For decades military thinking has been dominated by the idea that to threaten to strike back is the best way to 'deter aggression'. For nuclear weapons, this idea is understandable: to use nuclear weapons only to strike an aggressor on your own territory would be suicidal; if a nation or alliance seeks to preserve peace, the only use to which it can put nuclear weapons is to threaten retaliation.

The idea that threat of retaliation is the way to deter aggression spilled over into strategy for conventional forces, and is one reason why the two alliances – NATO and the Warsaw Pact – have maintained conventional forces which in structure are unstable and menacing: forces designed to strike back differ little from those designed for a surprise attack.

The notion that in the nuclear age aggression in Europe would end only in disaster and that stability and peace would best be furthered by going for defensive rather than offensive forces was put forward in the 1950s, 1960s and 1970s by a number of distinguished military heretics – including Sir Basil Liddell Hart, General von Bonin, Commandant Guy Brossollet, Horst Afheldt and others. It was also argued that technology offered possibilities for designing cost-effective defences. From these starting points, the idea of developing defensive forces was pursued in some Western countries, notably West Germany, under a variety of names, for example 'defensive

1

defence', 'non-offensive defence', 'non-threatening defence,' or 'non-provocative' defence. But for a long time such schemes remained a heresy, discussed in rather isolated national groups which had more influence on opposition parties than on governments. One reason for this was that there then seemed to be little or no prospect that the Soviet Union and the Warsaw Pact could be persuaded to change over to defensiveness. Discussion of how to implement such a policy – as distinct from its pure theory – was therefore focused on proposals for a one-sided change to defensiveness by NATO. Such proposals, however, were not easily sustainable. In the first place, if one side on its own changes its force structure and doctrine to defensiveness, the difficulties and costs it will face in maintaining security will be much greater than they will be if two sides do so together. Added to which, NATO has been numerically much weaker in conventional forces than the Warsaw Pact.

The position has been transformed since Gorbachev came to power. As a result of the 'new thinking' he introduced, defensive doctrine has been adopted by the Soviet Union and the Warsaw Pact, and, moreover, a reduction in the conventional forces of both sides to below present NATO levels has been accepted as a first objective in arms negotiations. It is in connection with this change that the chapters in this volume will be of particular interest to the reader. They have their origin in a series of meetings of a Workshop of the Study Group on Conventional Forces in Europe organised by the Pugwash Conferences on Science and World Affairs. The workshop met for the first time in Copenhagen, in 1984, and has since met once or twice a year in various parts of Europe. No less important than these meetings have been the contacts among its members which have permitted a growing informal exchange between East and West of ideas about defensive doctrine and conventional arms negotiations.

The purpose of the workshop at the start was to stimulate research into defensive defence and to bring the idea of defensiveness into current thinking about strategy and arms negotiations. This meant bringing together the Western analysts from different countries who had so far worked in relative isolation; opening a dialogue between East and West where previously there had been virtually none; and involving civilian analysts and military professionals in a common effort to develop valid concepts. At a time when contacts with the East were not easy to establish, Pugwash, with its long-established tradition of East–West dialogue on disarmament, was the obvious forum for such an exchange. Since the early meetings took place at the time when there seemed to be little or no hope of early flexibility on the side of the Warsaw Pact, the early papers were predominantly by Westerners and focused on the one-sided case, that is, the case where only one of two adversaries changes from an offensive to a defensive force structure. Apart from being 'politically realistic', this was the way to explore the possibilities of increasing defensive strength *vis-à-vis* a possible attack.

In the early stages, the work consisted of setting out and elaborating the general principles underlying the defensive approach, in examining specific proposals for possible defensive arrangements on the Central Front between NATO and the Warsaw Pact, and in developing methods for analysing the interaction between traditional military units and more defensive units. Analyses of the latter kind have been grouped here in Part II under the heading 'Combat Dynamics'.

While views differed on many practical points and many questions remained unsolved, it seemed clear that the basic idea was sound, and that the specialisation of forces for defence provided a hopeful and realistic path towards stable *détente* and arms reductions in Europe. As the Conference on Security and Cooperation in Europe at Stockholm was ending, and talks about the mandate for the next round were about to begin, the general principles underlying the defensive approach were set out in a memorandum which was circulated to the thirty-five delegations to the conference and discussed with a number of delegates at a meeting, in April 1986, in Stockholm. The memorandum suggested that further examination of the concepts and the ways of implementing them in practice would require, in addition to formal negotiation and informal exchanges among scientists, a dialogue with high-ranking military experts from both East and West and public debate on these issues.

By this time official policies were in flux. The members of the Warsaw Pact, meeting in Budapest in June 1986, publicly advocated a defensive doctrine: in an agreed statement they recognised the need to base the military concepts and doctrines of the two alliances on defensive principles. NATO replied, in December 1986, in a declaration which called for negotiations on conventional forces aimed at eliminating the potential for surprise attack and for large-scale offensive action. The new Soviet doctrine was not without ambiguity. It included the rather imprecise objective of 'Reasonable Sufficiency'. But subsequently Soviet statements made it clear that, as regards conventional forces, this meant forces sufficient for defence but unsuited for attack – a point spelled out by Gorbachev in a reply to a letter sent to him by four members of the workshop – Albrecht von Müller, Frank von Hippel and ourselves – in October 1987. Both sides were now beginning to prepare for new negotiations on conventional arms reductions in Europe which would be aimed at reducing the capability for attack on both sides, not merely at creating numerical balance.

These more promising prospects were reflected in the group's work and are reflected in the later chapters of this volume. Attention shifted from the question of designing new force structures for one-sided implementation to discussions of concrete proposals made by both sides for selectively reducing offensive capabilities of the two sides, in such a way as to achieve mutual defensive superiority. The group began to consider air forces and naval forces, possible arrangements for areas other than the Central Front, and possible approaches to negotiation. The chapters, which were written over a period of about five years, thus reflect a changing political background and the progress that has been made in that period in thinking about how to unwind the military confrontation by changing strategy. They should be read as a record of a cooperative analytical venture which is far from complete. The work continues.

Part I
Principles

Part 1

Principles

2 Deterrence and Defence
Anders Boserup

Virtually all thinking on armament and disarmament is based on the deterrence concept and, as many of us suspect, this may be one reason why disarmament efforts have been so barren. They have degenerated into arms control talks and into an accompaniment to the arms race, seemingly designed to legitimise and help it along rather than to impede it.

Such discussions as there have been were prompted, it appears, less by sound scepticism than by the desire to defend deterrence against the emerging nuclear war-fighting doctrines. These are indeed ghastly and should be opposed. But they cannot be counterposed to some 'pure' deterrence, as if these were alternatives. Deterrence is no escape from doctrines of, and preparations for, nuclear war-fighting. Rather, it is their foundation; they too are meant to deter. They are the logical outcome of attempts to preserve the credibility of the deterrent where it might be in doubt.

What we need is a hard look at the notion of deterrence itself; at its assumptions, at its implications for security and disarmament, and at the adaptations and extensions which the abstract armchair notion of deterrence must receive when it is implemented in practice. Instead, we have tended to eulogise a fictiously purified notion of deterrence, declaring 'abuses' and 'excesses' taken in its name to be unpalatable implications of the doctrine.

The basic problem in all nuclear deterrence is its inherent lack of credibility. This is because the reprisals envisaged are, by definition, wholly disproportionate in their effects and militarily pointless. While one can hope to deter hostile actions by threatening to massacre millions of innocent people, it makes no sense whatever, when the time comes, to carry out the threat, even if it can be done with impunity. For deterrence to be credible, not rational political judgement but insane vindictiveness or some unthinking 'doomsday' device must be seen to be in control. In asymmetric deterrence, where lesser forms of transgression are involved, the means to enhance credibility are even more destabilising. A plausible link must be established between the deterrent and the situations in which it is supposed to function. This can be done in various ways: providing the physical means for step-wise escalation; propounding with self-assurance

7

scenarios of survivable nuclear war; setting up more or less automatic tripwire arrangements; fostering such intense confrontation that a commitment to suicide and massacre for limited goals carries conviction, and so forth.

I do not deny that deterrence can work. I only wish to call attention to its darker side, to the requirements which must be met if it is to be more than a paper tiger. These requirements are but the 'excesses' referred to above: uncompromising hostility and paranoia, recurrent sabre rattling, and military and political dispositions which promote instability and encourage escalation. Brinkmanship, meaning the deliberate creation of credible avenues of escalation in crisis and in war, is the essence of deterrence. If brinkmanship is too timid there is no credibility, hence no deterrence. If it is too vigorous there is war. Deterrence must be dangerous, genuinely dangerous, if it is to be.

It follows that there can be no such thing as a 'stable', 'pure' or 'minimal' deterrent. Either it would not deter or it would not remain stable, pure or minimal for long. In fact, stability in deterrence is a self-contradictory notion; each of its terms can be achieved only at the expense of the other. Nor is it obvious that a minimum deterrent would promote security and stability in the world. Returning to concepts akin to the doctrine of massive retaliation of the 1950s is more likely to promote confrontation, brinkmanship and insecurity because credibility in real-life situations would be low. If we had a minimum deterrent, sensible people would soon be calling for more 'flexible' and 'graduated' capabilities to achieve deterrence without constantly courting disaster.

These attempts to salvage deterrence by dissociating its 'good and desirable elements' from its 'abuses' are futile and will only entangle us in self-deception and self-contradiction. Instead, it is in the direction of a sensible defence policy that a viable approach to peace, security and disarmament must be found.

First of all, the two concepts – deterrence and defence – must be kept apart. Very often they are not, the term 'deterrence' being used to cover both ideas, denoting any approach designed to prevent war by military preparedness. This slipshod concept seems designed to promote self-righteousness and to obstruct analysis. Provided only that our motivations are pure, any forces of ours, however excessive, destabilising and provocative, can be referred to as 'our deterrent'. And they 'deter' not only enemies but war itself. The excellence of deterrence so conceived follows from its definition.

I propose instead to confine the term 'deterrence' to its proper meaning. It has the same Latin root as 'terror' and should denote a policy of dissuasion based on threatening reprisals which would outweigh any conceivable benefits from attack. 'Defence', on the other hand, is a policy of dissuasion based on counterposing such force that an attack would be certain to fail.

Both deterrence and defence serve the aim of dissuasion, and one is not more bellicose than the other. But it is essential to keep apart not only the concepts but the things they denote, for these two modes of dissuasion are incompatible in practice. If they are combined into one doctrine, as has happened for Europe, a frightful military mess arises, combining the worst aspects of both: insatiable and mutually stimulating demands for arms, sabre-rattling and deliberately escalation-promoting postures. The theatre nuclear force deployments in Europe are one recent example of this.

Doctrines combining deterrence and defence cannot avoid self-contradiction. The logic of deterrence and the logic of defence are each perfectly cogent, but military dispositions which enhance the capacity for defence erode, for that very reason, the credibility of deterrence. Generally speaking, every measure is beneficial for dissuasion from one point of view and is detrimental from the other – by how much no one knows. A doctrine mixing both kinds of logic is therefore irremediably ambiguous. If, in spite of this, it is used to decide on the best course of action, it assumes the logical structure of a self-contradictory theory. By arguing now in terms of deterrence, now in terms of defence, as suits your case, you can 'prove' any point whatever, 'legitimise' any military decision whatever. Yet it is precisely this kind of doctrine that has passed for serious strategic analysis for years and on which European 'security' is based. It should come as no surprise that military arrangements in Europe, with huge armies arrayed for action and dependent in defence – even in the frontline – on systems designed as tripwires for the ultimate deterrent, were conceived in an intellectual framework of this calibre.

It is generally assumed that military stability and some measure of security can be achieved through a balance of force. This has been the fundamental axiom in all disarmament negotiations to date. But it is utterly wrong. Stability does not arise from an equality of force but from an inequality: the superiority of defensive over offensive capabilities. It is essential to be clear and precise about this fact. The more obviously the defensive capabilities on both sides suffice to meet any contingency, the more perfect the stability. This is the essence of the concept of dissuasion through defence. If, contrariwise, offensive forces on both sides outstrip the defence capability of the opponent, no amount of careful balancing can create stability. The balancing of forces as ordinarily conceived is completely irrelevant for security and military stability; the decisive factor is the ratio of defensive to offensive strength.

For this same reason the adoption by states of a defence approach to security need not lead to an arms race. On the contrary, it can lead to disarmament, and it is probably the only viable approach to it. In deterrence, each is compelled to sustain and create sources of instability. In a policy based on the balancing of forces, each threatens the other by overinsuring on the basis of a worst-case analysis, thus engendering an arms race. Genuine security and arms restraint can only be based on the notion of preserving peace through forces which are ample in a defensive role but whose offensive capability is deliberately reduced to a minimum. Only in this way can I reconcile my demand for security with your right to it, and the promotion of global security through *détente* and disarmament with my primary concern: national security.

The idea is not new. The objection one meets is that it is impossible to distinguish in general, abstract terms between offensive and defensive weapons. That is indeed true, and it would no doubt be impossible to reach agreement on blanket prohibitions of specified types of weapons in disregard of the military, geographic and other circumstances of each case. Such an approach was tried in the League of Nations and, understandably, it failed. But when the total force structure of a country is considered in the concrete military–geographic context in which that country finds itself, it is often possible to identify force components which are perceived by others as potentially offensive and which could be replaced by more unambiguously defensive systems without loss of security.

Countries have no difficulty in identifying (in their opponents, of course) those force components which threaten stability and justify counter-armament. This being so, there is scope for countries to consider each other's apprehensions and to move towards greater reliance on purely defence-capable, instead of dual-capable, systems. Even modest steps in this direction would greatly enhance security and *détente* in the area concerned and thus facilitate disarmament. Moreover, there is no need to engage in complex negotiations. Unilateral steps would create the objective conditions for arms reductions on the other side, and even if there is no reciprocation, all will have gained in terms of enhanced security.

In areas such as Europe, nuclear weapons have been assigned the *military* role of assisting in an otherwise conventional type of war, including both defensive and offensive missions. In these cases a defence-only approach to security must imply efforts to develop adequate conventional defences without resort to nuclear weapons, even if this should mean more hardware and higher defence budgets. It is an essential step towards removing the incentives for the other side to develop nuclear war-fighting doctrines and capabilities. And it is gross irresponsibility to let conventional defences lapse if this means relying on a 'tactical' nuclear component instead.

Generally, in the situation in which we find ourselves in Europe, it ought to be a primary concern on both sides to avoid as far as possible force postures which the other side would find worrying, should they ever doubt our purely defensive intentions. Efforts to design our forces so that they do not offer worthwhile military targets for nuclear attack would further diminish the relevance of nuclear weapons for a prospective enemy. If the *military* use of nuclear weapons is rendered unnecessary for me and ineffective for my opponent, and if my force posture is such that I cannot pose a military threat to others, then the risk of nuclear war in Europe is effectively dispelled. An enemy who cannot succeed militarily might, of course, still engage in a gratuitous massacre of civilians. Some may regard scenarios of nuclear blackmail and lunacy as likely. I do not. But in any case, this is the only type of contingency for which a nuclear deterrent might still seem useful. That, briefly stated, is what I think a 'minimum' deterrent should be: a deterrent whose *role* is reduced to a minimum.[1]

Note

1. This chapter was previously published in *The Bulletin of the Atomic Scientists*, December 1981. It is reprinted by permission of the *Bulletin*, a magazine of science and world affairs. Copyright © 1981 is held by the Educational Foundation for Nuclear Science, 6042 South Kimbark Avenue, Chicago, Illinois 60637, USA.

3 Defence Dilemmas
Alexei Arbatov

INTRODUCTION

The large-scale measures for the unilateral reduction in the Soviet armed forces, announced by Mikhail Gorbachev at the United Nations, have sparked off discussions on questions of military doctrine among Soviet politicians, the military and scientists. The discussions centre on several key problems. But here I would like to dwell on two issues: first, how military doctrine correlates with offensive military operations, which can be carried out for defensive among other purposes; and secondly, on the interrelation between military parity and reasonable or defence sufficiency.

The task of averting war – both nuclear and conventional – proclaimed in the new Soviet military doctrine does not obviate the need to train and raise national armed forces for the performance of military operations should war be forced on the country by a potential enemy. What is at issue is what kind of operations Soviet troops must be ready to carry out. Revolutionary changes in Soviet military doctrine are clearly manifest, inasmuch as the clear-cut objective to prevent war on the basis of defence sufficiency entails a substantial overhaul of traditional strategy, operational plans, tactics and military constructions. Formerly, it was generally accepted that the greater the Soviet military potential, the stronger the country and the more stable security, Soviet and international alike. Now a different approach is needed. Ruling out attack by means of credible deterrence, this potential must not in itself engender the fear of likely aggression on our part, no matter how unthinkable such action would seem from the angle of our political intentions. The maintenance of this potential should be proportionally combined with arms reduction and limitation talks and not stand in the way of efforts to curb the arms race.

ATTACK AND DEFENCE

It is well known that even conventional modern weapons, to say nothing of nuclear arms, possess a tremendous destructive power, together with high velocity, long operational range, good mobility and sophisticated methods of control. All these facilities can be used for both offensive and defensive operations. The aggressor, however, has an intrinsic edge – he chooses the time and place of the attack, making it exceedingly difficult to withstand unflinchingly. True, it makes sense to say that no defence is feasible without planning offensive operations underpinned by the appropriate forces and facilities, which primarily take a counter-offensive form. At the level of strategic arms, a counter-offensive means a retaliatory strike which causes unacceptable damage to the perpetrator of the initial attack. In terms of conventional arms, mobile defence includes responding to incursions, counterstrikes, flank attacks and counter-offensives with the aim of driving out the aggressor and frustrating his plans.

But can offensive and defensive strategies be equalised? For example, as General Ivan Trebak has pointed out: 'As before, defence is for us the main form of military operation. It is obvious, however, that defence alone is unlikely to ensure the final rout of the enemy. Therefore troops must know how to carry out an offensive'. General Anatoly Gribkov holds the view that this 'does not run counter to the principles of military doctrine for local wars since, as is confirmed by the experience of the Great Patriotic War and local wars, such actions are not only possible but also necessary as part of defensive operations and battles in particular directions'. Acknowledging the importance and authority of the above statements, I would like nevertheless to raise a question. If defence is 'as before' the most important thing for us – if in the past as well as now the main emphasis was on defence – what then is new in our military doctrine and why should we change our defence-oriented strategy and military potential, as well as making profound unilateral cuts? Further, what does it mean 'to ensure the final rout of the enemy'? In a major war, does that mean to win? Certainly, defence alone will not achieve victory. But neither can victory now be achieved through offensive action. A war between the Soviet Union and the United States or between the Warsaw Pact and NATO could not be won at all: neither a world war nor a war in major theatres, neither a nuclear war nor a conventional war. In all cases the only possible outcome is overall catastrophe. What has this to do with victory or a 'final rout'?

Historical experience always lies at the foundation of the science and art of war. But it needs to be approached creatively and can only be used selectively. Due to our unpreparedness for war, – caused by the gross military and political miscalculations on the part of Stalin and his entourage, arising from their criminal repressions against the army and the people – German troops reached Moscow in the space of three months and the Volga a year later. Then, by means of counter-offensives on all fronts for a period of three years, we fought to win back the lost lands. Three months versus three years, losses many times greater than those of the enemy, the horrors of captivity for millions, occupation for tens of millions, devastated land . . . God spare us from repeating that experience! But the experience can serve to stop us repeating the same errors. This means in the first place that the front line must be backed up by a reliable defence that will not permit the enemy to advance deep into our territory; and that counter-offensive operations will consequently be less important. At the same time, I must admit that the counter-offensive remains relevant as an element of defence. What matters is the scale of planned counter-attack operations: the size, characteristics and the deployment of troops. This is probably the key dilemma for those designing defensive doctrine with regard both to nuclear and conventional arms: what must be the potential and aims of counter-offensive capabilities if they are to protect the country from aggression and at the same time not in themselves create a threat to other states?

For all the complexity of this issue, a reasonable distinction between defensive and offensive strategy nevertheless can and must be drawn. For example, when emphasis is placed on strikes by strategic forces against the other side's strategic weapons and control systems, using high-accuracy and high-capacity nuclear facilities with a short approach time, and there is emphasis on the well-synchronised timing of their strikes as well as on establishing defence systems to intercept the enemy's surviving weapons – all this points to an offensive strategy oriented primarily to delivering a first strike. If, however, attention centres on the ability to strike the enemy's administrative and industrial installations (which are relatively few), if emphasis is placed on the viability and preservation of strategic facilities and their control systems even at the expense of their destructive power and speed, all this points to the planning of a retaliatory strike and a defensive strategy.

With conventional forces, assessments must be made not only in terms of the number and quality of the weapons but also in terms of their deployment, structure and composition, as well as their reinforcement potential. At a tactical level, the same divisions, their arms and military hardware can carry out offensive and defensive operations, as both kinds of operation can be required for a defensive strategy. At the strategic level, however, it is quite possible to distinguish between offensive and defensive orientation. If the forces are stationed along the frontier, this points to a defensive strategy. If, however, the troops are arranged in striking 'fists' at specific points along the line, this could indicate offensive plans. Being aligned for attack does not necessarily mean advantages for defensive purposes if the enemy is also poised for the offensive. But a situation of this kind makes for instability, and intensifies the danger of an armed clash. From a defence perspective, it is most sensible to deploy troops along the front (or in threatened areas) along defensive lines, while keeping a strike reserve (the second line) in the rear for a counter-attack to stop a likely breakthrough and drive back the enemy. The stronger the defence on the front line and the more serious the steps the two alliances take to restructure their military potential on strictly defensive principles, the less each side needs a big military formation for counter-attack, and the further in the rear that formation can be stationed, generating less fear on the other side.

PARITY AND SUFFICIENCY

The interrelationship between parity (or 'balance') in military strength and defence sufficiency is another topic on the agenda of the current debates. It looks as if parity cannot be counterposed to reasonable sufficiency, nor can the two concepts be equated. If defence sufficiency is the lower limit of the country's military might safeguarding it from aggression, it is obvious that this limit has its own definite parameters and depends on the other side's strength. But, as some authorities on military affairs write, can the limit of reasonable sufficiency depend on the need to maintain military parity – presumably at the lower rather than the highest level? If parity is understood as an approximate balance in the forces of the two sides, it would make no sense to reduce the concept of sufficiency to the above concept.

Seeking parity obviously means striving to have our forces and therefore our military potential approximately equal to those of the opponent. The enemy is, however, guided by its own political ends, doctrine and strategic principles. This is fine if the ends pursued are of a defensive nature and the other side is sensible and is seeking the maximum possible arms reduction. But what if it is not? What happens in the latter case? Should we then reject sufficiency and be unreasonable too?

Sufficiency presumably implies a much higher degree of independence from the strength and the steps taken by the opponent, and an intention to pursue one's own political and strategic ends and resolve one's own problems at a practical level, naturally taking into account countervailing aims, concepts and strength. If for example, we intended at all costs to maintain approximate balance and parity with the US strategic forces, given the constantly increasing emphasis on a first strike that determines their levels and structure, it would objectively imply an emphasis on a first strike on our part as well. That would be wrong both from the point of view of our defensive strategy and in view of the fact that the main principle of Soviet military doctrine is to avert nuclear war.

Back in the 1960s experts estimated that 400 nuclear warheads with a yield of one megaton each are capable of causing unacceptable damage to the biggest country in the world, destroying up to 30 per cent of its population and 70 per cent of its industrial capacity. Defence sufficiency would mean our ability to do similar damage by means of a retaliatory strike even under the most adverse conditions of attack. So to the amount of unacceptable damage we have to add the volume of forces that could be hit by a US first strike and intercepted by its defence facilities in the most unfavourable conditions for us during an exchange of strikes. This is evidently the factor that determines the dependence of defence sufficiency on the enemy forces which is, of course, not tantamount to parity. If the enemy forces are much more efficient and one's own are unreliable and vulnerable, one could then have parity without sufficiency. Whereas if our tremendous national resources are spent effectively so as to produce a different force structure with the emphasis on quality, reliability and viability, then sufficiency can be achieved even with less potential, that is, it can be achieved without parity.

In the sphere of conventional arms, the criteria for defence sufficiency are, of course, more complex. Proceeding from the assumption that it is impossible to wage a prolonged conventional war between NATO and the Warsaw Pact in Europe (a war of more than a few weeks), defence sufficiency presumably envisages the prevention of a victory within a brief period of intensive war and the avoidance of any escalation of war to the nuclear level.

CONCLUSION

A discussion presupposes the airing of different views. Not all of them are equally competent and substantial, since the issues under consideration are exceedingly complex. In this connection I should like to draw attention to the statement in an article by Yuri Lebedev and Alexei Podboryozkin that discussions latterly 'have demonstrated insufficient knowledge by politologists of the military doctrines, and sometimes a predilection for hasty conclusions along with lack of professionalism. Indeed, this is partly explained by insufficient specialised knowledge and partly by the fact that the discussions involve such professions as journalists, specialists in related disciplines (economists, geographers, even linguists), writers with a somewhat hazy graph of the topic in question'. All that is correct, but the lecturing tone here is very wide of the mark. For many years our professionals kept any information as to the numerical size, composition and objectives of the Soviet armed forces under wraps and we did not know anything about their total budget or the purpose of this or that costly weapon system. Today, too, the situation with regard to *glasnost* in the military field leaves much to be desired. How could journalists (and 'linguists') possibly help being perplexed? After all, our well-informed colleagues have had a hand in this also. For example, for many years they held forth to the effect that the zero option in medium-range missiles would give NATO 'a double preponderance in carriers and a triple edge in warheads', that there was 'an approximate balance in tanks and armoured hardware' and that there was an overall parity in Europe in conventional arms. And now we are going to scrap thousands of tanks and guns, hundreds of aircraft, and still, I presume, not exceed NATO levels. Those who had to have a clear 'idea of the topic in question' were trying to prove that our military doctrine has always been and continues to be strictly defensive, that military construction was gauged in terms of levels of reasonable sufficiency, and that *perestroika* in the army boiled down to the eradication of 'grand-daddyism'. But we are not now reducing our potential by half a million men and 20 000 major weapons units and reviewing the systems of training commanding officers and principles of field exercises for troops to the detriment of our defence capacity, are we?

The list of these complaints could be continued were it not obvious that the lack of *glasnost* and democracy in defining military policy, the limitation of the participants to a handful of 'professionals' and 'those with sufficient specialised knowledge' have led our policy to a state of hypertrophied military methods of giving us security, as well as to excessive waste of resources and grave miscalculations. The only way to resolve these problems is to carry out a profound restructuring based on democratisation and *glasnost*. Here no one can claim a monopoly of the truth.[1]

Note

1. This chapter was first published in *New Times* (Moscow), June 1989.

4 Conventional Arms Control: The Agenda and its Dangers
Albrecht von Müller

INTRODUCTION

The political long-term goal is clear: a peaceful transformation of the East–West conflict through marginalisation of its military stratum. Even if it has become a dirty word to some people on the western side of the Atlantic, I will not shy away from referring to this as a 'second *détente*'; the difference from the first being that this time attention is focused not only on political and economic relations, but also on a true reduction of existing military threats in Europe. This could lead to real conflict transformation and *détente* being made irreversible.

Conventional arms control and arms reduction will, however, be an extremely difficult task. It is obvious that the subject matter is of a much higher complexity than in the case of nuclear arms control. The fact that instead of two major actors, we have a double digit number of nations which has to reach consensus makes early stalemate, if not even complete failure, a fairly safe bet. Despite all this, the creation of a military regime of structural stability lies truly in the interest of all European nations, including the Soviet Union – and it should be consistent with the enlightened self-interest of the United States as well. Therefore, it seems worthwhile to push hard for effective conventional arms control.

The purpose of this chapter is to spotlight the agenda and the major political and practical dangers and obstacles that lie ahead. Three main issues will be touched upon, namely, the need for a solid conceptual framework; the need to adapt the ongoing force modernisations to the goal of stability during the negotiating phase; and, finally, the need for fairly large steps and not too gradualistic an approach.

It should be stated at the outset that we will try to avoid the trap symbolised by the experience encountered in the Mutual Force Reduction (MFR) or Mutual Balanced Force Reduction (MBFR) Talks:

the focus should not be on bean-counting. All participants must understand from the very beginning of the process that two Guderian-type armies, even if they were exactly symmetrical in character, would not provide stability at all. Instead, emphasis must clearly be on a change of force characteristics rather than on purely quantitative reductions, since only the former can make the defensive capabilities of both sides clearly superior to the respective offensive capabilities. Only through such an intentional and systematic decoupling of offensive and defensive capabilities can structural stability be reached, and the inevitable margins of doubt and ambiguity in evaluating the other side's capabilities be rendered obsolete. What is really quintessential is that we reach a consensus about this specialisation on defensive missions and capabilities as the paradigm for the new round of arms control already on the agenda. Otherwise, we should continue the MBFR/MFR Talks as they are. But this brings us already to the first main topic – the need for a solid conceptual framework.

Obviously, there exist two different ways of reaching this goal of structural stability: one can selectively disarm the most threatening and offence-capable components, or one can focus on the defensive aspect through a specifically designed modernisation. In an ideal world, one would try to follow only the first path. In practice, we will have to find an intelligent combination of both approaches, though – beyond doubt – the selective disarmament approach is preferable. But the overall goal of the process is too important for us to allow it to be endangered by being too rigid about the approach.

THE NEED FOR A SOLID CONCEPTUAL FRAMEWORK

Reading Western as well as Eastern statements and position papers on conventional arms control, one clearly observes a conceptual transition period. The old paradigm of balance, parity or symmetry still comes up on many occasions. Therefore, it may be useful to stress once again the difference between the notions of symmetry and stability: balance and parity *per se* do not provide stability. Just think of the confrontation of two Guderian-type armies, already mentioned. If a large bonus can be gained from pre-emption and attack, as in a gun duel between two cowboys, great instability can result not only despite but often because of balance, parity or symmetry.

The only way to create a stable regime is to reduce these bonuses for attack and pre-emption as much as possible and to exploit fully the structural advantages of the defender. This is possible today, technically as well as militarily. The core argument can be quickly summarised. The main problem of the defender is information, because he has to compensate for the momentum of surprise and local superiority on the side of the attacker. But the technological changes of our epoch have brought and are bringing major progress, especially in the field of information-processing and information-gathering. Therefore, the overall trend is going against the mobile platforms, which are still crucial for the attacker today.

But even when this shift from symmetry or parity to stability has been intellectually digested, a lot of troubled waters lie ahead. What does it really mean that somebody possesses an attack and offence capability? How can we identify and separate offence-prone structures from those which provide an advantage for the defending side? In the early phase of the debate on alternative conventional force structures and strategies, the tank was seen as the main culprit. Today, we know that this was an oversimplification and that a lot of different weapon systems and their synergisms constitute attack and defence capabilities. Therefore, a much more detailed analysis is needed. To this end, it seems appropriate to divide what we mean by 'offence and attack capabilities' into three main categories.

First, there is *the pre-emptive component*. It comprises all the actions and weapon systems by which the military capabilities of the defensive side are reduced in the very first phase of a military conflict and before the defending side has really unfolded its own military capabilities. With regard to weapon systems, this threat mainly consists of three elements: rocket systems with a range of more than approximately 50 kilometres and small Circular Error Probable (CEP); attack aircraft with a relevant payload and sufficient penetration capabilities; and, last but not least, sabotage-type measures which can also play a crucial role in crippling the other side's defences.

The second category then relates to *the actual events on the battlefield*. Here the offence relies mainly on six elements, namely, artillery preparation by cannons and rockets, a quantitative and qualitative superiority of main battle tanks and other armoured vehicles, highly mobile logistics, foward-deployed ammunition stockpiles, bridging equipment, airborne troops and, to a certain extent, tactical air support.

The third category comprises various *methods and measures to weaken the civil and military infrastructures*, and thereby break down military and civil resistance. Here again, rockets with long ranges and small CEPs, air attacks and sabotage measures play a crucial role.

These three categories constitute the backbone of conventional attack and offence capabilities. It goes without saying that this is only a very rough scheme and not at all complete. But for our purposes it is enough because we will try to decouple offensive and defensive capabilities to such an extent that the unavoidable margins of ambiguity do not matter too much.

To conclude these reflections on the conceptual framework a last topic has to be touched upon – the form of reductions. There are basically three forms of reductions: equal constants, equal percentages or equal ceilings. As Figure 4.1 shows, the first type is very much in favour of the side that had superiority at the beginning. Instead of stabilising these reductions destabilise. The second approach does not worsen the situation in principle, but it does not improve it either. (If there exists something like an 'operative minimum', this approach may even lead to a severe deterioration of stability.) Only the third approach – to go down to equal ceilings – means real progress, though even this has to be complemented by changes in the characteristics of the forces as outlined above. The

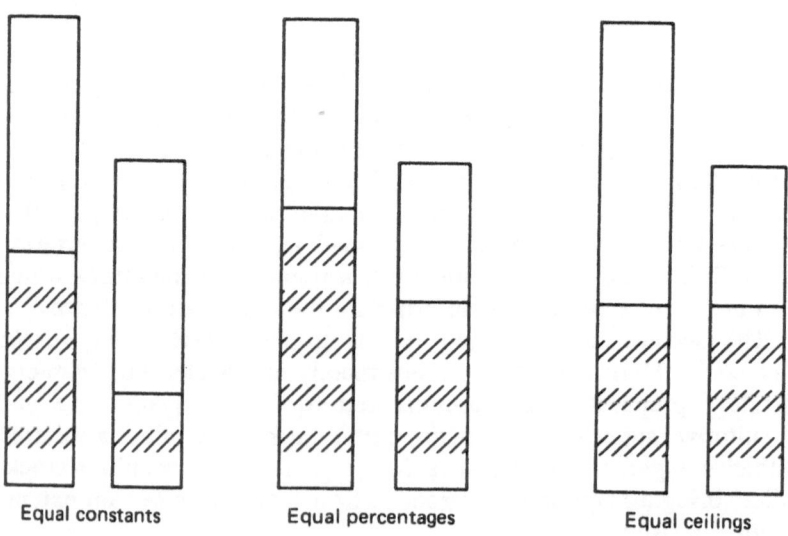

Equal constants   Equal percentages   Equal ceilings

Figure   4.1

third approach at the same time is the only one which reduces the bean-counting problem. As soon as one agrees upon ceilings for verifiable weapon systems, it no longer matters which side had how much at the beginning. Both sides simply have to reduce until they reach the agreed ceiling.

With respect to this ceiling approach, which can and probably should be applied individually to the major weapon categories, a warning to the West is necessary. Many clever analysts in NATO defence ministries and their back-up institutions tend to define ceilings as just about what the West has at present (see Figure 4.2a). I advocate equal ceilings, but there must also be some reductions by the side that was or felt inferior before. If, for example, *A* has 10 and *B* has 6 at the beginning, a ceiling of 6 would mean a 4:0 bargain against *A*. This is not really a thrilling offer. A ceiling of 4, instead, would not drastically increase the concession *A* has to make, but would imply that *B* also has to give up a relevant share of its former potential (see Figure 4.2b). Even if *A* really is committed to arms control, it cannot possibly accept the first version. But it can, and rightly so, sell the second solution at home as a real breakthrough in arms control. Yet it is predictable that many NATO defence ministry analysts will claim that NATO is already at its operative minimum

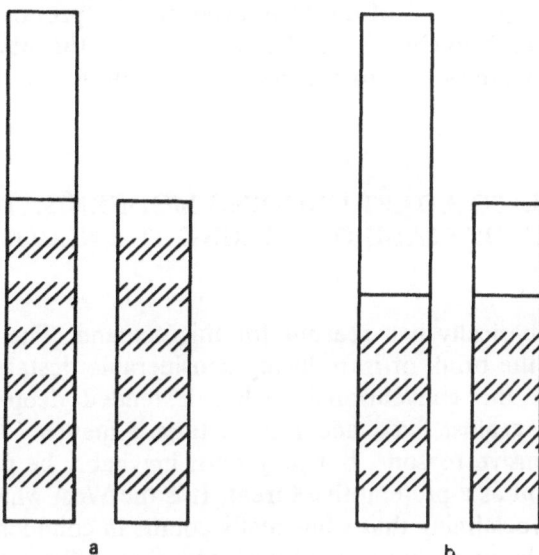

Figure 4.2

which it cannot go below, no matter what the other side does. But if
this were really to be the final position of defence ministries,
responsible politicians should seek somewhat brighter military
experts. If one wants to stabilise the European situation militarily, it
is impossible to stick to all aspects and features of present forces. But
it can be demonstrated that through selective disarmament steps and
a reorganisation of present force structures much stability can be
gained, not to speak of a clever use of modern technology.

When we ask what really might be attractive for Mikhail Gorbachev in conventional arms control, we receive two answers. One is
that he has a sincere interest in stabilising the European situation.
This may be the case and I personally tend to believe it. The second
answer is obvious and legitimate: he wishes to save money and free
techno-economic potential for his reforms. NATO should not allow
this occasion to pass by. It should come up as soon as possible with
a fairly drastic conventional arms control proposal that includes
relevant reductions on its side also. The Treaty on Intermediate
Nuclear Forces (INF) is the best evidence that Gorbachev is ready to
make significantly larger concessions where the Warsaw Pact has
previously enjoyed superiority. But it goes without saying that this
approach has to be applied also to those (very few) weapon categories
where NATO is presently ahead. But we will come back to these
questions of pace and of concrete arms control proposals after we
have touched upon the second issue, namely, the need to adapt
ongoing modernisations to the goal of stability during the phase of
negotiating.

THE NEED TO ADAPT ONGOING MODERNISATIONS TO THE GOAL OF STABILITY DURING THE NEGOTIATING PHASE

There are basically two reasons for this demand. The first is that
we are on the brink of introducing considerable destabilising technologies in the conventional realm. Extended deep-interdiction
capabilities against fixed and mobile targets may be acquired for
purely defensive reasons, but they must inevitably be perceived by
the opponent as a pre-emptive threat. It is the West which for years
has been proclaiming that what really counts in confidence building
are not declarations but capabilities. And that is absolutely correct,
but exactly for this reason deep-strike capabilities of either side are

devastating for crisis stability. They directly increase the bonus for pre-emption which is inversely proportional to crisis stability. If we fail to pay attention to this problem, we might easily end up in a situation in which destabilising effects take place which by far outweigh the positive effects agreed upon at the negotiating table. Therefore, we should learn from nuclear arms control and should try to make sure, if possible in the consensus about the agenda, that no further destabilising modernisations take place during the phase of negotiating.

Before turning to the character of such an agreement, it is necessary to touch upon the second need for such an initial agreement. It is what might be called the 'bargaining chip trap'. Again, we know from experience with nuclear arms control that many extremely questionable procurement programmes survived only thanks to the argument that they at least would result in bargaining chips, and that giving them up would weaken the negotiating position. Many critics, therefore, argue that arms control has often been perverted into the best way of legitimising very questionable procurement programmes, thus accelerating rather than slowing down the arms race. When we think about an intelligent design for conventional arms control, we should be aware of this trap and try to avoid it from the very beginning. How can this be done? The astonishing answer is: apply almost unchanged the basic maxim of Immanuel Kant's theory of ethics, namely, the categorical imperative, to the security and defence issues we are considering. It reads like this: 'Defend yourself by those and only by those means and capabilities which would not worry you, if the other side possessed them, too.' This is probably the quintessential maxim for improving conventional stability. Its application requires in principle two things: first, ongoing and planned modernisations should be announced and discussed openly; and secondly, either side should, ideally, have a veto against specific developments – though making use of this veto would mean that one also binds oneself to accept the same treatment.

So much for the principles. We would, of course, face some additional problems in our concrete historical situation due to existing asymmetries. If we apply these principles without restriction, the side which is ahead in a specific field would have a strong incentive to veto further developments, thus perpetuating its existing superiority. Therefore, it would be necessary to introduce – and this may sound very strange at first glance – areas of procurement programmes that would be 'protected' against being vetoed. As attack and defence capabilities depend largely on being able to bring mobile platforms into the other side's territory, one might agree that platform-stopping weapon-systems, which at the same time by their very structure cannot be transported into foreign territory, should be the 'protected' procurement area. At the same time, one should try to restrict strike capabilities, which can be misused for a crippling pre-emption, as soon as possible.

In summing up, we can say that we are dealing here with extremely tricky problems. But at the same time the attempt to avoid the drawbacks of nuclear arms control is worthwhile. Probably we will not find a completely satisfactory solution for the problem of adapting ongoing modernisations to the goal of stability. But without even trying to do so we would surely be worse off; and we might end up in a situation in which during and maybe even due to arms control negotiations conventional stability continues to deteriorate rapidly.

THE NEED FOR LARGE STEPS

Complex social systems do not make quantum leaps. In the case of economic changes, as well as other reforms, normally only a fairly gradualistic approach has a chance of achieving positive results. The only exception to this general rule, known to the present writer, is the field of arms negotiations – the reason no doubt being that it is in the first place a very strange thing to negotiate with someone about the nuclear and/or conventional kill mechanisms targeted against oneself. The complexity of 'conventional' warfare means that the effects on combat capabilities of minor corrections and changes are extremely hard to evaluate. For example, if there was a proposal to get rid of 10 to 15 per cent of Warsaw Pact artillery, some Western experts would be sure to argue, counter-intuitively, that this would not reduce but might even increase the offensive options of the Warsaw Pact because the slimmed-down forces would be more flexible and mobile than those that went before.

This argument, which explains the need for rather large steps in terms of the complexity of conventional warfare, is probably the most important one. If we do not want to be trapped into a new form of bean-counting, now applied not only to weapon systems but also to military options, we definitely have to go for fairly large steps. And there exist two additional arguments that go in the same direction. They are of minor relevance but should not be neglected.

The first argument is that only fairly large steps make it possible for the West also to give up something. It is not completely incorrect when Western analysts stress time and again that NATO is at a sort of 'operative minimum'. Their proposal that in the field of conventional arms control only the Warsaw Pact would have to dismantle weapon systems and units is based on this perception. But such an approach is simply not negotiable. As mentioned previously, the INF Treaty provides the best evidence that Gorbachev is prepared to give up more than NATO in fields in which the Warsaw Pact is clearly ahead. But we should not expect that he will propose to his political decision-making bodies, and especially to the Soviet military establishment, an agreement which implies that only they have to give up things, while the Western side may even have the right to increase certain weapon systems. As stated already, if this were the Western approach, it would be better not to start conventional arms control negotiations at all. As well as not leading us anywhere, it would necessarily worsen political relations between East and West. When negotiating with someone, it often makes sense to assess his motives. Why should the Soviet Union be interested in arms control agreements which will definitely result in asymmetric reductions, mainly to their disadvantage? The answer is not entirely simple. But, according to my assessment, Gorbachev understood that he could not win the arms race against the West and, even beyond that, that the arms race is sort of a 'synthetic' or 'artificial' problem, and that it is high time to get down to the real tasks of our epoch, such as the preservation of our global environment, the reduction of the North–South asymmetries, and so forth. It is possible that the fairly strong interest in arms control results from just such an overall interpretation of the East–West conflict. In addition, this is probably backed up by a strong interest in saving money in order to gain flexibility for economic and social reforms. And here we come to conventional arms control because in the Soviet Union nuclear forces consume only a minor fraction of what is spent on conventional forces. Given the correctness of this supposition, a strong interest on the part of the Soviet Union in conventional arms control becomes plausible, as does even the preparedness to give up an existing edge in the conventional realm. But, as we have noted, nobody can go home with arms control agreements which provide disarmament steps only for himself.

Another and probably less important argument for large steps relates to the policy style of Gorbachev. It must be assumed that proposals for relevant conventional disarmament steps will not only run into bureaucratic inertia but will raise active resistance from the military services. And in this respect it has been Gorbachev's method to make clear-cut decisions and achieve definitive results very quickly before the momentum of resistance and opposition can build up. No doubt this effect of meeting less resistance when 'cruel decisions' (cruel for specific societal interest groups) are made quickly also pertains to the Western side. It is impossible to imagine that something like the Reykjavik proposals could have been the result of a policy-formulating process of ministries and agencies. It takes a powerful and personally committed decision-maker (or a very poorly informed one) to get to those radical solutions even before the status-quo-defending immune systems start to work. (Seen from the Soviet viewpoint, it was probably a major mistake not to accept the Reykjavik package, even without an agreement on the Strategic Defense Initiative (SDI). After President Ronald Reagan's retirement the SDI programme and much related research will fade away anyway due to increasing recognition of technological obstacles and fiscal restraints.)

It seems appropriate to end this chapter with a rough draft of what could be a stability oriented arms control regime for Europe. The proposal outlined below is clearly painful for the Western side, but it is undoubtedly more painful for the Warsaw Pact. Its main advantage is that it could not be refused by any democratic government in the West that wants to be re-elected. (With a certain time-lag this eventually should be true for France.) The substantial provisions might be roughly grouped around the following three issues:

● The first element should be fairly low ceilings for the backbone components required for a strategic offensive and the conquest of territory, that is, main battle tanks, artillery and Multiple-launch Rocket Systems (MLRSs), armed helicopters and land combat aircraft. Both for main battle tanks, as well as for heavy artillery/MRLSs, ceilings of approximately 10 000 might be sensible. These ceilings would have to be combined with a maximum density limit that prohibits attack-capable concentrations. Here one could conceive of a maximum density of 0.02 per square kilometre measured with a grid width of 50 × 50 kilometres. A sensible ceiling for armed helicopters might be approximately 500 and for land combat aircraft it should not exceed 1000. With these ceilings the mobility required for an efficient defence is still guaranteed, while at the same time the mobility and redundancy that is required for strategic offence is denied.

• A second element should be limits on forward deployed ammunition stockpiles, mobile logistics and bridging equipment. These also have the purpose of reducing the mobility factor to the degree and characteristics required for defence. The restriction of forward deployed ammunition stockpiles also severely impedes attack, since the attacker must prepare a breakthrough by extremely concentrated shelling.

• The third element should be strict limitations regarding the ranges, as well as the numbers and the deployment patterns, for all those weapon systems that might be misused for a crippling pre-emptive strike. Here one is mainly referring to rocket systems, but one might also have to consider such systems as artillery with extended ranges and Remotely Piloted Vehicles (RPVs). A careful analysis of crisis stability and pre-emption bonuses shows that the most influential single factor proves to be range in combination with the CEP of these weapon systems. If one or even both sides are capable of executing a crippling conventional first strike, it is impossible to regain stability by any means. Simulations show that up to a range of approximately 50 kilometres the advantage of being able to concentrate precision fire is mainly on the side of the defender. Beyond that – and especially in 'deep interdiction' with ranges of several hundred kilometres – it becomes an extremely destabilising factor, and constitutes overwhelming escalatory dynamics towards military action in crisis.

This proposal intentionally does not provide limitations on typical 'platform-stopping' equipment such as passive ammunition. It therefore combines selective arms reductions with a rechannelling of such future investment in arms as is unavoidable. Altogether it provides an autogenous stability for the conventional realm. This would pave the way for a reduction in the military role of nuclear weapons and for the nuclear disarmament of European territory, as well as for further reductions of the remaining offence-capable components in the future. At the same time, Europe will not be completely denuclearised as sea-based systems do remain. Given the low yields and small CEPs of today's Submarine-launched Ballistic Missiles (SLBMs), this implies that NATO could even stick to its overall doctrine of 'flexible response'. (This is a major advantage because any arms control proposal that would require a new NATO strategy to be developed and agreed upon would mean a time loss of at least a decade.)

Thus the proposed nuclear-conventional deal has something really attractive in it for both sides:

- The West gets rid of its conventional inferiority and an autogenous stability for the conventional realm is provided. This should make it possible for the West to accept a far-reaching disarmament process and the reduction of the nuclear component to sea-based systems.
- On the other hand, the Soviet Union and its allies achieve drastic reductions of military expenditures in combination with the long-desired nuclear disarmament of Europe. This should be attractive enough to make the Warsaw Pact accept drastic cuts in its conventional forces and the loss of offensive options that hitherto existed, at least according to the Western assessment.

As to the long-term political implications of such an arms control regime, it is clear that increasing the responsibility of the Europeans for their own security, and reducing the military role of both superpowers in Europe, must not of necessity weaken the alliances. It should, on the contrary, strengthen those alliances which are based on a joint political value system. To sum up: such a stabilisation of Europe, in combination with a partial unwinding of the arms race, should give more room and flexibilty for the political dynamic of a peaceful transformation of the East–West conflict.[1]

Note

1.   This chapter was submitted to the 37th Pugwash Conference on Science and World Affairs held at Gmunden am Traunsee, Austria, 1–6 September, 1987.

5 The Confrontation of Conventional Forces in the Context of Ensuring Strategic Stability

Andrey Kokoshin and Valentin Larionov

In May 1987, at the Berlin Conference the Political Consultative Committee adopted the document entitled, 'On the Military Doctrine of the Warsaw Pact Member States'. The principal substance and orientation of this military doctrine are those of preventing both nuclear and conventional war, and safeguarding and defending socialist countries from outside encroachments. The cutting edge of the Warsaw Pact doctrine is directed not at preparing for war, but against war and at consolidating the principles of international security. It is the first time a provision on the prevention of war has been included in a definition of the Warsaw Pact's military doctrine in such a pointed manner. The Warsaw Pact's military activity envisaged a struggle to prevent war in the past, too, but this task has now moved to the fore in the doctrine, and has become the main and definitive task.[1]

The conclusion that victory is impossible in a nuclear war and the task of preventing both nuclear and conventional war have also been stressed in the joint statements of the Soviet and American principal leaders in Geneva, Reykjavik and Washington. All this makes us take a new look at certain traditional, long-established approaches to the implementation of purely military tasks.

Many eminent political and military figures, specialists and academics in the West have increasingly, especially in recent times, been coming out with declarations about the need for the military activity of the opposing military alliances to be of an exclusively defensive orientation. The question of the United States, France and Great Britain following the example of the Soviet Union and the People's Republic of China in assuming an obligation of no first use of nuclear weapons has been raised more than once in the West. There is more

31

active criticism of destabilising operational concepts: 'Air-Land Battle' and 'strikes against second echelons and reserves' (the 'Rogers Plan'); the US naval strategy formulated for the 1980s, which is linked to the names of former US Secretary of the Navy John Lehman and Admiral James Watkins; and other concepts. Moderate political circles, in which an increasingly active part is being taken by high-ranking military professionals, are persistently proposing and working on so-called alternative military ideas and concepts of 'unprovocative defence', 'non-offensive defence', 'non-offensive deterrence', and so on.

One of today's central issues is how the political aims of preventing war and strengthening strategic stability are reflected in the military–technological part of military doctrine: in strategic operational concepts; in the deployment of armed forces; in mobilisation plans for industry, and so on; and of how the transformation and development of the military-technological part of military doctrines will be conducted in the process of arms limitation and disarmament.

This chapter examines in sequence four hypothetical options for the two alliances' confrontation at the level of conventional forces and weapons. Each of these options is fairly theoretical and schematic, and is offered as material to stimulate research. In many of their parameters the nature of confrontation and the criteria and conditions for ensuring peace in this sphere are considerably different from those existing at the level of strategic nuclear forces. At the same time, it should always be remembered that the conventional forces of the two alliances are highly saturated with the nuclear weapons which their ground, air, and naval forces have at their own disposal. This gives rise to a need for a radical solution to the problem of reducing the levels of weapons which are dual-capable systems.[2] It stands to reason that these options far from exhaust the conceivable forms for this confrontation. It is possible, for instance, to imagine various combinations of them. However, the schemes developed here can be regarded as one of the analytical instruments for progress in strengthening strategic stability in Europe and in relations between the Warsaw Pact and the North Atlantic Treaty Organisation (NATO).

The essence of the first option is that each bloc is oriented towards immediate counteraction – strategic offensive operations – if a war should begin (an attack by the other side). In this context military operations would be of a decisive and uncompromising nature. It can be assumed that in this situation each side would strive to transfer combat operations to enemy territory and air space as rapidly as possible, in order not to subject its own territory to excessive devastation and radioactive contamination, among other reasons.

This option corresponds to the deep-rooted tradition of military thought according to which only decisive offensive operations and efforts to take the strategic initiative will lead to victory, and victory in turn lies in the final rout and destruction of the enemy's forces. This tradition assumed its clearest form as far back as the nineteenth century, during the Napoleonic Wars, and dominated military and political thinking during both world wars of this century.

Offensive strategic and operational concepts have always seemed psychologically attractive in a number of ways. Many people consider them necessary to keep up the armed forces' morale today, too. Such ideas have traditionally been popular among a considerable proportion of the public which has a poor grasp of the realities of military affairs.[3]

From the start of military operations, this kind of confrontation to all intents and purposes orients the sides toward a series of meeting engagements. These are traditionally regarded as one of the most complicated forms of military operation, and one which demands a particularly high degree of skill in troop management. Under contemporary conditions the complexity of managing operations of this kind is increasing to a limitless extent. A readiness to conduct such engagements introduces additional tension into the military–political situation and increases mutual suspicion. The fact is that this orientation demands maximal combat readiness and that large troop exercises and headquarters training exercises be constantly conducted. This confrontation option also presupposes an appropriate structure of forces and weapons composed of groups prepared for meeting and counter-meeting engagements. In this context it is an exceptionally complex matter to distinguish what is intended for pre-emptive offensive operations from a defensive position, and what is intended only to rebuff an attack. Reconnaissance–strike missiles (guided-missiles) and fire-control complexes may become one of the principal means of waging armed conflict in the future. In these complexes the launch and kill are carried out within the range of fire (up to several hundred kilometres) through the guidance (homing) of missiles and warheads over their entire trajectory, or some sectors of it. The control and operation of the systems are based on wide application of radio-electronics, laser technology, and computers: they are highly sensitive to jamming and vulnerable to the fragmentation and blast effect of munitions. Timely destruction of the corresponding headquarters communications networks and control centres will make it possible to reduce sharply the effective use of such systems. Because of the speed of operation of these systems, strikes must be inflicted on them immediately after the characteristic signs of reconnaissance have been detected. In other words, among all the measures which would prevent the operation of these complexes, a decisive role will be played by pre-emptive nuclear and conventional destruction, by the capture (or neutralisation) and electronic suppression of automated control systems, as well as by the interdiction of reconnaissance activities.[4]

Second operational echelons and reserves in the rear have a substantial role to play in this option. This in turn implies that second echelons will also be the targets of powerful long-range strikes from the earliest stage in the development of an armed conflict by means of aviation, surface-to-surface missiles, and also the future reconnaissance–strike systems. The battle may thus spread far beyond the lines of immediate contact.

In this confrontation option, the start of an armed conflict will immediately put large numbers of troops into action, and very complicated combinations will occur as they regroup. It would be extremely difficult to return to the original situation and restore peace. This means that one should assume in advance that the military conflict will be irreversible and that its intensity and scale will grow, while allowing for inevitable difficulties and even for the possibility that the political leadership and higher military command will lose control.

The transient nature of combat operations, the changes in the operational situation, the simultaneous involvement of large areas of European countries in combat operations, the intentional disruption of communication channels, and the conduct of combat operation at any time of day in any weather conditions will quite probably lead to a situation where the political leadership and higher military command will be prevented from keeping events fully under control, due to a lack of time and information. In extreme cases, this could take the shape of an irreversible escalation of military operations, up to and including the use of tactical nuclear weapons. The transition from combat operations where only conventional weapons are used to operations involving weapons of mass destruction may be sudden and unforeseeable, and this leads to a desire to keep one's nuclear weapons in increased combat readiness, which in turn substantially increases the danger that a nuclear war will start and escalate.[5]

A second option is that each side orients its strategy and operational art towards the deliberate renunciation of the offensive in the initial stage of a conflict, with the intention of conducting only defensive operations. In this case the emphasis is placed on a deeply echeloned, well-engineered positional defence, and on previously prepared forces for the counter-offensive. After the offensive has been repulsed in a defensive battle, in which a retreat and the abandonment of some part of one's territory are allowed for, there is still a capability to use reserves drawn from the rear in order to move over to a decisive counter-offensive (if necessary, even a general offensive), up to and including the complete rout of the enemy on his territory. Counter-offensive operations may be conducted both at the operational level (armies, fronts, flotillas, and fleets) and on a strategic scale (groups of fronts and fleets in the theatre of military operations).

The basic ideas and schemes behind such a defence, the principles of its formation and engineering preparation, the system of distributing forces and weapons, and the nature of the defensive battle can be seen in general terms in the example and historical experience of the Battle of Kursk in the summer of 1943.[6] It stands to reason that no direct analogies can be drawn between the Battle of Kursk and the forms of a confrontation in contemporary conditions. The organisation of the defence at Kursk took place during wartime and had completely different political and military–strategic motives from the primarily non-offensive structure of armed forces and their strategies, and from peacetime operational plans which are aimed at preventing war.

In assessing the pros and cons of the second option for the confrontation of the two blocs, it may be noted that from the point of view of strengthening strategic and international political stability it is interesting above all because of its idea of deliberate defence. As far as the actual nature of operational troop formations and the engineering preparation of defensive lines are concerned, all this must be made the subject of more detailed comparative research and joint discussion by representatives of the two alliances. The research and discussions could also examine the degree of dispersal of the defence, the disposition of forces in depth, the nature of positional defence combined with its activeness, and so on.

This option would seem to be more stable than the previous one.

Its shortcoming is the complexity of distinguishing and of monitoring each side's capabilities (preparations) for counter-offensive operations and for pre-emptive offensive operations, although the difference between the two would nevertheless be more obvious than in the first option.

With this option, the probability of a conventional war going nuclear is just as high as in the first option, especially if the sides retain a capability for counter-offensive operations which turn into a general offensive. The complexity of control and monitoring of the development of events by the higher political and military leadership would also be very considerable in this option, although monitoring would be easier than in the first option.

A third option envisages that the sides are capable only of routing an invading enemy formation on the territory they are defending without going over to a counter-offensive outside their borders. The essence of this defence is that combat operations are not carried on to the territory of the side which has begun the war. The result of the defender's active operations is simply the restoration of the situation that existed before the start of military operations (*status quo ante*). The two sides' possibilities of conducting active operations are limited, on a mutual basis, to the operational scale, that is, to a capacity for counterstrikes at army group or army level. Accordingly, the concept of victory is admissible only on the operational and tactical scales, but is ruled out on a strategic scale.

Allowing for a number of assumptions, an analogy for such operations can be found in military history in the rout of the grouping of Japanese troops which carried out the aggression against Mongolia in the Khalkhin-Gol area, in 1939. Conducting a brilliant operation to surround the invading troops and cut them off from the state border, the First Army Group of Soviet–Mongolian troops under the command of Corps Commander G. K. Zhukov utterly defeated the 6th Japanese Army in the Mongolian Desert between 20 and 31 August 1939.[7] A very important contribution to this victory was the achievement by the Soviet Air Force of air superiority in the course of this operation, for at the beginning of the armed conflict the Japanese possessed air superiority. The aggressor was taught a harsh lesson. However, no invasion of the territory from which the aggression had been launched was undertaken, although there were definite, purely military opportunities for a 'retaliatory operation'. The enemy requested a truce. Taking account of all the factors which defined the military–political situation both in Asia and Europe, the Soviet leadership made a decision not to respond with military operations on the Chinese territory occupied by Japan. Hence, on 16 September 1939, military operations ceased as a result of negotiations between the Soviet Union and Japan. In the final analysis the rout of the Japanese in the Khalkhin-Gol area substantially stabilised the situation in the region.

The Korean War, or rather one of its phases, may serve as another relatively recent example. This concerns the tacit agreement not to cross a certain boundary or demarcation line and not to extend the scope of military operations in the fourth phase of the war, after the situation had stabilised. As is well known, the war in Korea was the first major local war after 1945. From 10 June 1951 to 27 July 1953, until the signing of an armistice in Panmunjom, combat operations between the sides were of the nature of mobile defence, with sporadic battles in regions adjoining the 38th Parallel. The limited scale of combat operations was expressed in the fact that by tacit agreement the two sides refrained from steps which might rekindle a major conflagration. The troops of the People's Republic of Korea and the Chinese national volunteers did not penetrate deep beyond the 38th Parallel into South Korean territory, and American aviation refrained from bombing targets in the territory of the People's Republic of China.

Under contemporary conditions the difficulty in implementing this defence option lies in defining the size of the territory which may be lost, together with the wealth situated there, and whether each side will or will not agree to respond merely by restoring the *status quo ante*, suppressing its thirst for vengeance. The second point is how to measure the scale of compensation for losses suffered by the attacked side. The third issue is when this side will allow itself to stop: at the precise moment when the invading group has been eliminated, or will it go further? In the region of the Khalkhin-Gol river, the Mongolian territory invaded by the Japanese was desert country, and there were virtually no losses among the civilian population or material wealth during the fighting.[8]

A fourth option is for each side to choose, on an agreed basis or on the basis of mutual example, a purely defensive stance at the strategic and operational levels, without having the material potential for conducting offensive or counter-offensive operations.[9] High mobility would be mutually stipulated only for tactical troop formations which could be used for counter-attacks. This would mean a battalion, a regiment, or, as a maximum, a division.

This kind of grouping obviously must not possess strike aviation or weapons which are effective in a surprise attack (such as reconnaissance–strike systems) or which have great mobility and striking power (tank and air assault divisions), and it must not have forces and weapons for deep strikes at its disposal. In short, such a defence could become an entirely 'non-offensive defence', although two of the three options set out above possess this characteristic to some extent. Accordingly, the concept of victory exists only on a tactical scale in this option. At the strategic and operational levels the concept of victory is ruled out.

When examining the above options, as well as all other options for confrontation in the sphere of conventional forces and weapons, it is necessary to take full account of the conditions in which combat operations might unfold, and also of the entire range of their consequences. When conventional weapons are used on a massive scale it is impossible to rule out a deliberate or accidental attack on the enemy's nuclear and chemical weapons, including nuclear and chemical munition stores, launchers, storage vehicles for transporting nuclear warheads and shells, loader/transporters, and other such targets. The consequences of this might prove to be the equivalent of using the corresponding means of mass destruction, which would upset the balance in tactical nuclear weapons and give rise to unforeseeable counter-actions. Under certain conditions strikes by conventional weapons are also capable of destroying the numerous European nuclear power stations and power installations. The result would, to all intents and purposes, be the same as that of an attack involving nuclear weapons, and the consequences would be considerably more tangible than those of the Chernobyl accident.[10]

As a whole, industrially developed countries, on whose territories there are many enterprises using the latest technological processes, may be more vulnerable in a military conflict. This primarily applies to Europe, the place of confrontation for the Warsaw Pact and NATO. The destruction of a large chemical enterprise or oil storage tank even by conventional weapons would lead to explosions and fires comparable in total heat release to a medium-sized nuclear explosion. The release of chemically or biologically active substances into the atmosphere after the destruction of industrial enterprises could have the same consequences for the surrounding population as the use of chemical or biological weapons. The destruction of nuclear power stations by conventional weapons would be equivalent to the use of radiological weapons.[11]

Comparing all four options for the confrontation of the two alliances at the level of conventional forces and weapons, one may conclude that the first option is the least stable. The fourth option corresponds most closely to the idea of strengthening strategic stability and reducing both alliances' military potential to a level of sufficiency, dictated only by the need for a defence which is taken to its logical conclusion.

One of the real problems of strengthening strategic stability is that the task of preventing nuclear and conventional war must obviously not be restricted to the period before the start of actual combat operations – 'D-Day'. Elements which would contribute to localising an armed conflict, which might arise if political and diplomatic 'safety devices' fail to work, must be thoroughly considered and made part of the mechanism of mutual relations between the opposing sides. The task is as complicated as it is necessary. Once again, the fourth of the options under discussion corresponds most closely to this requirement.

The burden of localising an armed conflict is particularly complex with regard to the problem of using tactical battlefield weapons. On the one hand, concepts of 'limited' nuclear war are destabilising, and on the other hand, it is dangerous to forego mechanisms which might prevent a one-off use of nuclear weapons (including an unauthorised use) from growing into an uncontrollable total nuclear war. In this respect the danger inherent in the presence of a large number of nuclear weapons organically linked to conventional forces should be stressed once again. Therefore, it is exceptionally important to make organisational and technical provisions for the higher political leadership and military commanders of both alliances to have complete control over nuclear weapons.

Proceeding from the belief that it is necessary to renounce rigid schemes to resolve the problems of preventing war (for example, those that offer either complete peace or catastrophic war), we should recognise the importance of seeking mutually advantageous agreements with every positive step considered useful. The complexity of implementing a genuinely non-offensive defence must not rule out other, less stable confrontation options which nevertheless represent movement towards a maximally stable balance.

All these four defence options can only be used for strategic analysis in cases of confrontation where there is a line of contact, or where there is some stipulated demilitarised corridor on land. As far as forces and weapons situated deep within the territory of countries and alliances are concerned, there are other parameters which must be taken into account. For example, the parties could develop principles for confrontation on the ground which take account of all ground forces to a stipulated depth, not just the forces grouped along the border and oriented towards accomplishing combat tasks. However, regardless of what events occur on the ground (hypothetically there could even be total peace there), an aggressor would be capable of carrying out a sudden strike from the air, via outer space, or from the sea. Thus, issues of lending a non-offensive nature to naval forces, air forces, and space weapons must also be resolved according to the principles of common and mutual security.

The proposals contained in the documents of the conference of the Political Consultative Committee of the Warsaw Pact held in Budapest, in June 1986, are *de facto* moving in the direction of implementing the fourth option. In accordance with these documents, it is proposed that a system for reducing armed forces and conventional weapons be developed such that the process of reduction would lead to a decrease in the danger of sudden attack and would correspond to the consolidation of strategic stability on the European continent. In order to achieve this it is proposed that agreement be reached on a significant reduction in the tactical strike aircraft of both military–political alliances in Europe right from the start, as well as on a decrease in the troop concentrations along the lines of contact. With the same goal in mind, additional measures would also be developed and implemented to strengthen the confidence of Warsaw Pact and NATO countries, and of all European states, that sudden offensive operations would not be undertaken against them.

It is necessary to be aware of the fact that the transition of both sides to the non-offensive defence option is bound up with very considerable difficulties. It will require an unprecedentedly frank discussion and joint resolution of many purely military issues which are becoming political issues before our very eyes. These include the evaluation of what is to be regarded as defensive, the basis on which the operational and combat training of the troops is to be conducted, and so on.

The exclusion of any of the existing types of armed forces and weapons from the future defensive military structure of the two alliances poses an entire range of serious questions of a political, military technological, and doctrinal nature, among others.[12] Defining the defensive and offensive potentials of formations, combined units, and units is no simple matter. It is an exceptionally complicated matter to develop some single criterion for comparing the quantitative and qualitative parameters of armed forces and weapons. There are, however, a number of pointers and characteristics which make it possible to assign a particular system or weapon to the primarily offensive or primarily defensive category.[13] In addition, the two sides could agree on some maximum quantitative level, anything above which would be regarded as a deviation from the defensive concept and an acquisition of offensive capabilities.

Tactical technical characteristics represent the most important of the qualitative indicators of a weapon system which make it possible to differentiate, within certain margins, between offensive and defensive weapons. The most important qualitative characteristics can be taken to include the speed, mobility, and multirole capability of a system, the extent of its protection and invulnerability, its provision with all-weather and night-fighting equipment, and so on. The relative correlation of offensive and defensive characteristics could be determined and agreed on by the two sides for every major weapon system.

These examples are evidence of the fact that even weapon systems with different basic purposes do possess certain common qualitative features, but that it is quite realistic to differentiate between their defensive and offensive capabilities. The main difficulty here is presented by weapon systems with multirole capability. This problem can be resolved by taking a weapon system's qualitative parameters in conjunction with other factors, such as the nature of deployment of forces, the distance from the frontline, and the number of systems in service. Even the most clear-cut offensive system becomes less dangerous when taken in conjunction with such limiting factors as small numbers and deployment far from the site of possible combat operations.

For military specialists the most important and sometimes decisive factor defining the nature and result of combat operations is the problem of personnel: the number of servicemen, the standard of their training, their morale, the skill of commanders, and also the organisational structure of armed forces, the strength level of combat formations, mobilisation plans for force deployment, the composition of reserves, and the methods of training servicemen. All these factors can be examined with respect to the potential for offence and defence.

A reduction in the number of major military exercises and

manoeuvres and, even better, their abandonment, could serve as graphic evidence of the two sides' defensive intentions. The same goal would be served by an increase in the required period of advance notification and by limitation on the number of participants in exercises which do not require advance notification.[14]

In the near future new accords in this sphere could cover the limitation of exercises in terms of troop, air force, and naval force numbers; in terms of their purpose and operational scale; and in terms of the size and situation of the regions where they are conducted. Subsequently, agreement should be reached to reduce exercises to the tactical scale. In this context they should be conducted to show defensive operations such as the holding of positions, and the inflicting of counterstrikes and counter-attacks.

One of the elements in the transition to a non-offensive defence could be the transfer of regular formations to reserve status.

The issues outlined above require detailed and thorough work by military and civilian specialists on both sides, always by reference to the criteria and the conditions required for military-strategic stability.[15]

Notes

1. D. Yazov, *On Guard Over Socialism and Peace* (Moscow, 1987), p. 27.
2. A. Arbatov, 'Deep Cuts in Strategic Weapons', *Morvaya Ekonomika I Mezhdunarodnye Otnosheniya*, no. 4 (1988), p. 18.
3. A. Svechin, *History of the Military Art* (Moscow, 1922), part 1, pp. 31, 46.
4. M. A. Gareyev, *M. V. Frunze: Military Theoretician* (Moscow, 1985) pp. 244–5; Yu. Molostov, 'Defence Against High-Precision Weapons', *Voyennyy Vestnik*, no. 2 (1987), pp. 83–4.
5. V. Shabanov, 'Conventional War: New Dangers', *Novoye Vremya*, 14 September 1986, p. 8.
6. A. Kokoshin and V. Larionov, 'The Battle of Kursk in Light of Contemporary Defensive Doctrine', *Mirovaya Ekonomika I Mezhdunarodnye Otnosheniya*, no. 8 (1987), pp. 32–40.
7. An idea of the scale of combat operations can be gained from the composition of sides' forces as of 20 August 1939. The Soviet–Mongolian troops numbered 57 000 men with 542 guns, 498 tanks, 385 armoured cars and 515 aircraft; the Japanese forces had 75 000 men, 500 guns, 182 tanks, over 300 aircraft, and no armoured cars. The Japanese side thus possessed superiority in terms of personnel, but

were at a disadvantage as regards combat equipment, especially armoured vehicles.

8. In his memoirs Zhukov remarked on the absence of civilian population in the area of combat operation as a factor which made reconnaissance work more difficult. See G. K. Zhukov, *Reminiscences and Reflections* (Moscow, 1984), vol. 1, p. 204.

9. A number of eminent military specialists from West European countries (including West Germany) consider that in conditions where a balance was established between the Warsaw Pact and NATO and a reduction in tension took place in the international situation, the West could well afford to take major unilateral steps towards a transition to purely defensive operational–strategic options in the hope that these would call forth analogous steps from the Warsaw Pact.

10. *Novoye Vremya*, 14 September 1986, p. 8.

11. L. Feoktisiov, 'Constructive Actions Aimed at Halting the Arms Race and Averting the Threat of Nuclear War', *Mir Nauki*, no. 3–4 (1987), p. 19.

12. Thus the renunciation of nuclear weapons (above all tactical nuclear systems) casts doubts on the effectiveness of the entire strategy of 'flexible response' which NATO adopted as far back as 1967.

13. A number of Soviet and Western experts consider that the primarily defensive weapons include anti-tank guided missiles, mobile surface-to-air missile systems, mines and explosives, various defensive fortifications, tractor-drawn artillery systems, and low-speed combat support aircraft without in-flight refuelling equipment.

14. In contemporary conditions, it is becoming increasingly difficult to make the training environment approximate to the combat environment; as the development of forms and methods of operational and combat training shows, the difference between training and combat increases with the improvement of the weapons of war and the growth of their destructive power.

15. This chapter first appeared in *World Economy and International Relations* (Moscow) (in Russian), no. 6 (1988).

Part II
Combat Dynamics

Part II
Combat Dynamics

6 Conventional Stability and Arms Control

Albrecht von Müller

What is meant by conventional stability and, how can it be improved? I shall attempt to answer these two problems, using a new methodological procedure that I and some of my colleagues have developed in Starnberg, West Germany. This makes use of Force Ratio Development Functions (FDFs). Changes in military force relations in the course of a conflict form the basis of our approach.

We can describe the strength of one side, let us say the defender's side, as a function of time (see Figure 6.1). Here we see in the starting phase, in which the pre-emptive and surprise advantages operate fully for the attacker, that the defender's military strength decreases quite a lot in a relatively short time. Later on, we see the medium phase of the conflict in which by now also the defender has fully developed his forces and, finally, we see the end phase of the conflict in which it is decided whether the attack is rebuffed or whether the defender is completely worn down according to the self-accelerating Lanchester effects. If we want to plot not only the strength of side A but also the strength of side B, we could utilise a three-dimensional graph in which we have two axes for A and B and one for time (see Figure 6.1). But since the following topological considerations are much more difficult to demonstrate with three-dimensional graphs, we will restrict ourselves to two-dimensional projections in which we have only the strength of A and B as the axes. (Figuratively speaking, this means we turn around our three-dimensional object in a way that the time axis runs exactly into the depth and thus seems to disappear.)

Utilising these simplified graphs, we can now draw a lot of relevant conclusions if we start from an initial force ratio and then plot the two possible Force Ratio Development Functions (FDFs): one for the case that side A is the attacker, the other for the case that side B is attacking (see Figure 6.2). Here we now recognise the four basic variants, of which the first represents utmost instability, because the attacking side finally wins the war. The fact that one or the other side wins is described in our graph by a crossing of the FDF with one of the coordinates. This means that one side still has some military

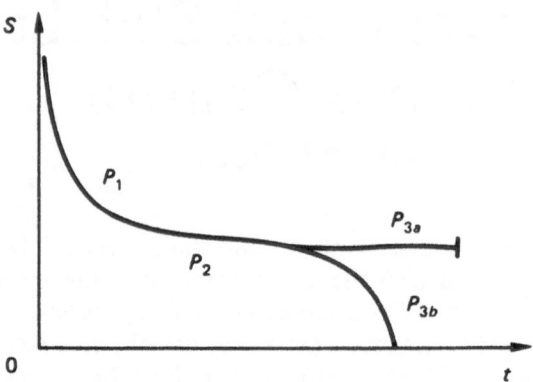

S   Strength of the defender
P_1 Opening phase with overproportional losses of the defender due to effects
    of pre-emption
P_2 Mid phase of conflict with fully developed forces on both sides
P_{3a} End phase with stabilisation of defender and breakdown of offence
P_{3b} End phase with accelerated breakdown of the defence
t   Time

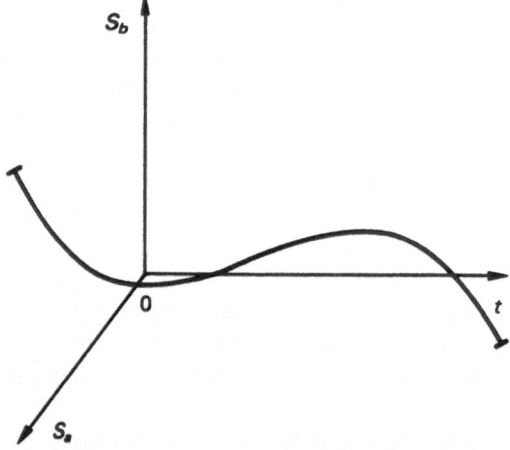

S_a Strength of side A; S_b Strength of side B.
Figure 6.1   The topology of 'Force Ratio' Development Functions as an
               indicator of crisis stability and arms control stability

forces left, while the other side is worn down completely. In the
second case, we now see a situation in which side *A* is absolutely
superior: no matter who attacks, in the end side *B* still has forces and
side *A* has none. The third case illustrates the exact reverse: side *A*

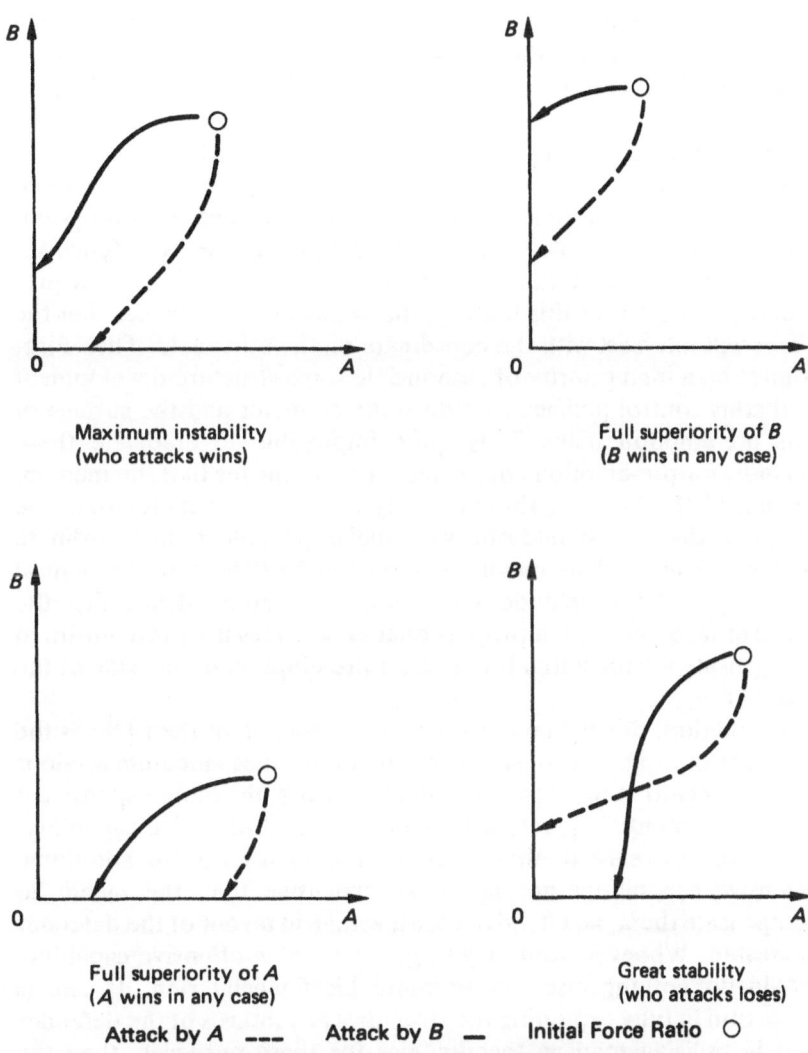

Figure 6.2 The four basic stability variants

enjoys absolute superiority. Those force ratios of absolute superiority are less disastrous for crisis stability, but they are nevertheless very uncomfortable politically because obviously the superior side can put political pressure on the inferior side. Therefore, both situations tend to lead to an arms race, because the inferior side will try by all means to become at least equal. The fourth and last case shows a really stable situation, namely, a force relationship in which the attacker finally loses the war. Topologically, this means that the FDFs intersect before they cross the axes. In this case, we have a situation of structural stability or of 'mutual defensive superiority', because the defence capabilities of both sides are clearly superior to the offensive capabilities of the respective opponent.

Building upon these fundamental characteristics of stable force ratios in the conventional realm, we can now deduce several optimisation criteria. The belly at the beginning of the two functions (directly behind the initial force ratio) represents the bonus for pre-emption. The bigger this belly is, the worse are the chances that the FDFs will intersect with the coordinate on the other side. Therefore, it must be a high priority of responsible force structure development and arms control policies to reduce the diameter and the surface of this pre-emption belly. It is quite impossible to eradicate these bonuses for pre-emption completely and this means that the medium section of the function should in any case be a bit flatter than the diagonal, that is, it should run as parallel as possible to the coordinate of the defender. This means that on the battlefield the structural advantages of the defender's side are fully exploited and that the attacker faces more than proportional losses. This is a prerequisite to compensate for the initial bonuses for pre-emption on the side of the attacker.

In addition, this flatness of the medium section of the FDFs is the indicator of arms control stability. The flatter these medium sections are with regard to the respective attacker's axis, the more expensive it is for the potential aggressor to gain additional offensive capability. But if this is realised, that is, if the marginal costs for additional offensive capabilities are far more expensive than the means to compensate them, an effective cost-leverage in favour of the defender is created. Whoever would try to gain a reliable offensive capability would do nothing else but promote his financial ruin. If one is successful in fully exploiting the structural advantages of the defender and in utilising modern technologies for these purposes, then the arms control process is no longer completely dependent on the good will of the other side. Through force structures themselves 'sane incentives' are created which favour restriction with regard to the build-up of offensive capabilities.

An additional interesting aspect of these topological analyses relates to the Janus-face of arms control. We see that through well-intended symmetric reduction it is possible to destabilise a situation which has been fairly stable before (see Figure 6.3). In this case all participants, being of good will, might fulfill their disarmament pledges without cheating, and still, in the end – due to unfavourably chosen structures – find themselves in a situation that would be far less stable than before.

On the other hand, a sensible and stability-increasing arms control policy requires that first of all the characteristics of the FDFs be changed in such a way that, as described above, the pre-emption belly is reduced and the FDFs are parallel with the defender's coordinates. Only if such a decoupling of offensive and defensive capabilities has taken place, can one start to reduce the armament levels without harming stability (see Figure 6.3). By means of this procedure the criteria most relevant to improving crisis and arms control stability can be precisely defined. First, the diameter and surface of the 'pre-emption belly' must be reduced as much as possible. Secondly, the middle part of the functions must, as far as possible, run parallel to

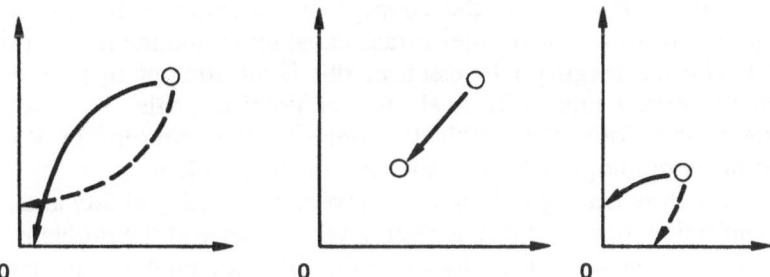

Negative case 1: Destabilisation despite balanced reductions

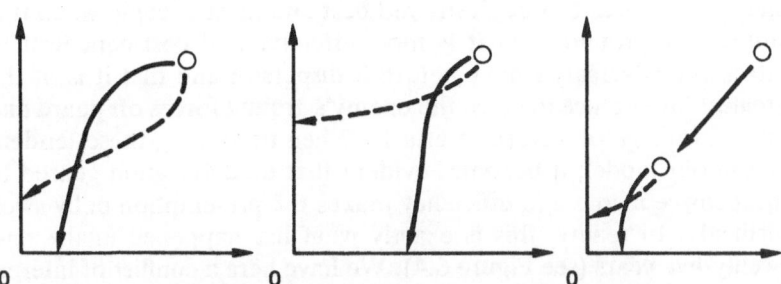

Positive case 2: Stability at a lowered level of armament

Figure 6.3   The arms control dichotomy

the attacker's axis. Finally, both functions must intersect in the area
of positive values while leaving a certain margin of security between
the intersection point and zero. If this topology of FDFs is guaranteed, then the highest degree of military crisis stability is reached.
Arms control stability is realised inasmuch as the middle part of the
function runs parallel to the defender's axis. The implication here is
that the defender is more efficient in combat and a cost-leverage
works against the marginal increase of offensive capabilities.

Let us next consider the widely discussed concept of 'structural
defensiveness' (*Strukturelle Nichtangriffsfähigkeit*). In 1982 and 1983,
this concept was developed at the Max Planck Institutite for Physics
and Astrophysics in West Germany and was conceived at the outset
as being exclusively for internal scientific use. The concept focused on
the problem of how to strengthen conventional defence, increase
crisis stability and improve chances for arms control at the same time.
(This is not to be confused with Horst Afheldt's concept of a purely
'defensive defence'. For the concept of 'structural defensiveness'
explicitly provides for counter-attack capabilities and the restoration
of territorial integrity.) In essence, this is an attempt to gear our
military means more effectively to our political goals. To a large
extent, therefore, the political critique of that concept in West
Germany did not touch upon the core of the problem.

But, before leaving the topic of conventional crisis stability and its
optimisation, one must not neglect to take a glance at the problem of
a possible conflict of aims between the drive for military efficiency
and crisis stability. When politicians charge the military with the task
of strongly enhancing the prospects for military success with as little
money as possible, the cheapest solutions will always have an
inherent pre-emptive tendency. In this vein it is supposed that the
enemy's air force can be destroyed best and most cheaply when it is
still on the ground; that it is most effective and cost-beneficial to
attack the adversary's navy before it disperses; and that it is of the
greatest importance to catch the enemy's ground forces off guard and
take advantage of a surprise attack. When translating these tenden-
cies in our model, it becomes evident that modernisation geared to
inexpensive increase in efficiency makes the pre-emption belly swell
distinctly. In reality, this is exactly what has happened in the past
twenty-five years (see Figure 6.4). We have here a conflict of interest
between the improvement of military efficiency with little money, on
the one hand, and better crisis stability, on the other. If better
integration of political and military goals and requirements is not

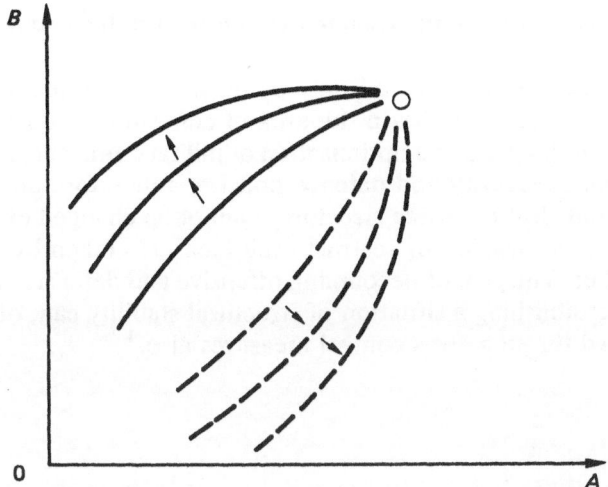

If military efficiency is to be increased inexpensively, measures
are often time-critical and have an inherent pre-emptive tendency.

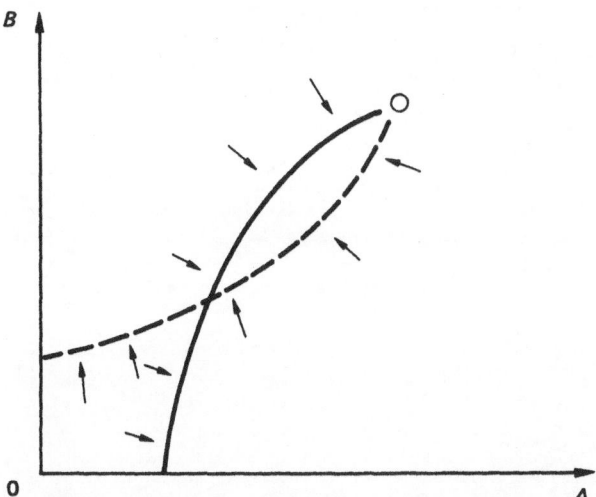

If crisis stability and arms control stability are to be increased,
pre-emption bonuses must be reduced and offensive/defensive
capabilities decoupled.

Figure 6.4 The decisive conflict of interest: increased military efficiency
versus increased crisis stability

reached, we will be on the road to ever more unstable and dangerous situations.

This trend can be reversed only by paying more attention to the problems of crisis stability in the area of conventional defence, and not through short-sighted optimisation of military efficiency. A better integration of security and defence policies is therefore an essential requirement. But the force structures cannot be changed exclusively through modernisation or restructuring measures taken by one side or the other. The goal of decoupling offensive and defensive capabilities and establishing a situation of structural stability can, of course, be pursued through arms control measures also.[1]

Note

1.   This chapter is a shortened version of a paper circulated to the Pugwash Study Group on Conventional Forces in Europe in 1987.

7 The Implications of the Increased Accuracy of Non-nuclear Weapons

Robert Neild

Some people say that, with the advent of precision-guided munitions, defence is coming into the ascendant again in non-nuclear war. Others say that the use of manned platforms to carry weapons cannot last much longer:[1] tanks, aircraft and warships will all become obsolete because they will be so vulnerable to future generations of precision-guided munitions. These statements may prove to be right. But it is hard to be sure. They are often *ad hoc* propositions by reference to particular vehicles (usually tanks) and particular weapons designed to destroy those vehicles (anti-tank weapons). They are vulnerable to *ad hoc* counter-propositions and, consequently, to untidy, inconclusive argument. For example, the experience of the 1973 Middle East War is often cited as if that war, more than ten years ago, provided a definitive demonstration of the new technologies and the possible strategies for their exploitation. The purpose of this chapter is to describe a general principle that illuminates the relationship between accuracy and defence.[2] The principle also shows why, when accuracy is high, it may pay the defender to disperse his forces rather than concentrate them.

The subject whether increasing accuracy favours dispersion or concentration, defence or offence, merits some consideration. We have recently been reminded of the failure, before the First World War, to recognise that advances in firepower, notably the development of the machine gun, had brought the defence into the ascendant. The conventional wisdom in 1914 was that the attack would be supreme.[3] In retrospect, that seems astonishing. It is tempting to reflect that, had the supremacy of the defence been foreseen, perhaps war would have been avoided. Perhaps the massacre in the trenches, caused by the supremacy of the defence, would have been avoided. Who can say?

Technology is changing again. Whether it redounds, or can be made to redound, to the advantage of the defender or attacker is a question we should try to get right. The only way to find out is by theoretical analysis of the implications of changing technology and by examining evidence yielded by simulations, field exercises and war. What is offered here is a tentative attempt at theoretical analysis.

It is convenient to start by considering a simple model of the exchange of fire between two forces using weapons that are *not* highly accurate. The model applies equally well to any exchange of fire where there is repeated shooting by one force against another in conditions that are symmetrical as regards cover, the absence of surprise, and other variables. For example, it can be applied to an artillery duel, a naval engagement, or an infantry engagement. What matters is that there is an open 'shoot-out' in symmetrical conditions.

We shall call our numerical example an infantry engagement. Suppose two forces of soldiers, Blue and Red, start shooting at each other in symmetrical conditions (for example, with the same rifles and standards of aiming) so that the hit probability is the same on each side. The only thing that is assymetrical is the number of men.

Suppose there are twice as many Blues as Reds. At the first round of fire (supposing they start firing simultaneously) the Reds can aim at only *half* the Blues. But on the other side *two* Blues can aim at each Red. The result is that the Reds, who were fewer to start with, lose nearly twice as many men as the Blues. Consequently, at the second round of shooting the ratio of Blues to Reds will be even more favourable to the Blues than at the first round. There will be only enough Reds to aim at less than half the Blues; and there will be more than two Blues to aim at each Red. The difference in casualty rates will, therefore, be greater than at the first round, and the ratio of Blues to Reds will show a further and bigger change in favour of the Blues. In this way, the differential advantage to the Blues will accelerate dramatically as shooting continues.

To construct the mathematical example shown in Table 7.1, I make these assumptions:

1. 2000 Blues fight 1000 Reds. (Large numbers are used so as to avoid fractions of a man.)
2. The kill probability per shot (k) is 0.2 on both sides and is assumed in this simple example to be the same as the hit probability, that is, all hits are fatal.
3. At each round all the Blues and Reds shoot simultaneously and the bullets cross in mid air. (In order to isolate and examine the consequences of numerical superiority and nothing else, we rule

out the possibility that one side or the other may gain by shooting first; and we rule out the possibility that the numerically superior side may stagger their fire within each round.)

4. Each man on each side at each round is given a target, selected on the basis that fire is allocated evenly to targets so as to maximise total kills. (If it is assumed that each man chooses his own target and target selection is random, the rate of kills will be less than in this example.)

Table 7.1   Simultaneous fire between Blues and Reds, $k = 0.2$

Round	Blues		Reds		Ratio of Blues to Reds
	Number at beginning of round	Losses during round	Number at beginning of round	Losses during round	Ratio at beginning of round
1	2000	200	1000	360	2:1
2	1800	128	640	297	2.8:1
3	1672	69	343	227	4.9:1
4	1603	23	116	111	13.8:1
5	1580		5		316.0:1

The figures can be explained as follows:

(a) In Round 1, 1000 Reds aim at 1000 Blues and kill $1000 \times 0.2 = 200$ Blues. The number of Blues surviving at the beginning of Round 2 is $2000 - 200 = 1800$.

(b) In Round 1 on the other side, 2000 Blues aim at 1000 Reds, spreading their fire evenly so that they aim two shots at each Red. Consider their shots as two waves, even though they are, by assumption, simultaneous. The first wave of 1000 shots will hit and kill $1000 \times 0.2 = 200$ Reds. The second wave will *hit* $1000 \times 0.2 = 200$ Reds. But because 200 of 1000 Reds will have been killed by the first wave, the number *killed* by the second wave will be only $800 \times 0.2 = 160$. So the Red losses in Round 1 are $200 + 160 = 360$. The number of Reds surviving at the beginning of Round 2 is $1000 - 360 = 640$. The ratio of Blues to Reds at the beginning of the second round is 1800 to 640 or 2.8 to 1.

(c) In Rounds 2, 3, 4, and 5 the process continues with the ratio of Blues to Reds taking off dramatically. Every shift in that ratio means fewer Blues will be shot at and that more shots will be fired at each surviving Red. These are two sides of the same coin – the coin which generates the accelerating change in the ratio.

Whether this law is applied to an artillery duel, a tank battle or other forms of exchange, the same conclusion follows, namely, that it pays to concentrate your forces for the sake of numerical superiority.

This is a principle that was discovered seventy years ago by Lanchester, an English engineer and inventor.[4] He called it 'the Principle of Concentration'. Along with it he put forward his well-known 'Square Law'.

It is impossible not to admire the beauty of Lanchester's Principle of Concentration. It provided a brilliant rationale for concentrating your forces – and dividing those of the enemy if possible – so as to achieve superiority of firepower, something which military men knew from experience to be of the greatest importance.

But the law depends on two implicit assumptions. The first and more obvious is that the advantages of being on the defensive, with dug-in positions, knowledge of the ground and so on, are assumed away, as are the advantages the attacker may enjoy if he achieves surprise.

The second implicit assumption, whose significance is less immediately obvious but which is of interest today, is that the kill probability of the weapons is fairly low, so that repetitive shooting is needed for an assured hit.

Suppose the kill probability was one, meaning that every shot was sure to destroy its target. There would be no point in numerical superiority in an exchange of fire – as distinct from superiority in total forces available to a nation or commander. The Blues, with their first shots, would kill 1000 Reds. There would be no point in having more Blues to kill each Red more than once. One thousand Reds would simultaneously kill 1000 Blues with their first shots, if the bullets of the two sides crossed in mid air. Of course, that would not happen. The fact that it would not happen is a key point; it brings us to a conclusion: *as kill probabilities rise, the value of shooting first in an engagement increases.* Clearly, it is no longer safe, as it is in a world of low accuracy, for the analyst, let alone the soldier, to assume that it does not matter who shoots first or what the sequence of firing is. Gone, long since, are the days of the Battle of Fontenoy (1745) when the French, we are told, said: 'Gentlemen of the English Guard, fire first.'

There is a second effect of higher kill probabilities. They mean that the rate of change in the ratio of forces in favour of the superior increases – and does so dramatically. In Table 7.1 above, where k (the kill probability per shot) is assumed to be 0.2, the ratio of Blues to Reds at the beginning of Round 3 is 4.9:1. If, instead, k is assumed to be 0.5, the ratio at the beginning of Round 3 is 344:1.

We shall for the moment assume that there is a kill probability of one for weapons with single heads aimed at single targets (i.e. aimed at one man or one weapon). We shall temporarily leave aside weapons which characteristically will kill more than one man or weapon, a category that includes projectiles with multiple aimed sub-munitions as well as weapons that cause destruction over an area, whether by blast, splinter, fire, chemical contamination or nuclear reaction. For a single-target weapon the maximum value of k, the kill probability, specified in terms of numbers of men or weapons killed is one. For the multiple target weapon there is no general upper limit.

The assumption that the kill probability for single-target weapons is one has these implications for an engagement:

1. Who sees first, by eye or sensor, and shoots first, kills.
2. Therefore concealment, from the eye or sensor, pays.
3. If dispersion helps concealment, as it will in most geographical settings, dispersion pays.

Thus we are led in a few steps to the opposite conclusion from Lanchester. In his low accuracy world, concentration pays. In our perfect accuracy world, dispersion pays.

In mathematical terms, we could write that the chance of being killed (which is accuracy seen from the receiving end) was a function of two probabilities, the probability of being detected (P_d) and the probability of being killed once you had been detected (P_k). Thus probability of being killed $= P_d P_k$, and chances of survival $= 1 - P_d P_k$.

Lanchester implicitly assumes that there is no problem of detection, $P_d = 1$, and the chance of survival in an encounter depends on $1 - P_k$. I am explicitly assuming here that $P_k = 1$, and the chance of survival in an encounter depends on $1 - P_d$. To minimise P_d, it pays to disperse.

How far P_d can be reduced will depend on the possibilities for concealment and protection afforded by the terrain and by the man-made structures and diggings on it. We know that improvements in the technical means of observation, command and control are making it easier to detect and identify targets and aim from far at them. In other words, P_d is rising – except where the improvement in technical means of detection is countered by better concealment, achieved by dispersion or other means.

We turn now to the question whether rising accuracy (carrying with it a rising kill probability) redounds to the benefit of the attacker or the defender. We shall consider first the question whether attacker or defender gains from the premium on the first shot, then whether attacker or defender gains from the high reward to numerical superiority.

Concealment and protection help the defender to reap the benefit of the first-shot premium; lack of concealment and protection help the attacker to reap that benefit. The point can best be expounded by considering extreme cases.

Suppose first there is no possibility of concealment or protection for a combination of two reasons:

1. The ground is bare and flat and there is no man-made cover or protection.
2. Surveillance systems are such that each side can observe the men and weapons of the other side and direct accurate fire at them.

In these conditions, which amount to a confrontation on a concrete desert, or a confrontation at sea with surface vessels, whoever shoots first can pick off his opponent's forces with his opening shots. In the extreme theoretical case, where the kill probability of all weapons was one and each side had weapons in sufficient quantity and of sufficient range to be able to cover all his opponent's forces, whichever side shot first would eliminate all his opponent's forces; he would wipe them out in his opening salvo; there would be a huge advantage to firing first: in other words, a great advantage to the attacker. The case is analogous to the argument in the nuclear realm that, if there are vulnerable weapon systems, for example land-based missile silos, there is an advantage to shooting first and a temptation to attack pre-emptively. It is a temptation that may become acute and difficult, if not impossible, to resist as regards surface ships and airfields.

Consider now the opposite conditions, namely, that there are rich possibilities of concealment and protection because:

(a) The terrain (rural and urban) offers much cover; and/or
(b) Surveillance systems are of limited capability.

In these conditions the attacker cannot pick off his opponent before he advances. He must advance and seek him out. In advancing, he will reveal himself or his vehicle to the defender. The defender, who has stayed concealed, can then be the first to fire and pick off his opponent. He gains the first-shot premium.

Of course, the attacker may try to conceal himself by the use of smoke or electronic counter-measures to upset communications and sensors. But anything the attacker can do in that line of business the defender can do too. There is symmetry. But as regards revealing yourself in order to advance, there is asymmetry. Whoever is on the attack – or counter-attack— must reveal himself in a manner that is not necessary to the defender.

This proposition applies to an engagement with good cover/limited surveillance, up to the moment of the defender's first shot. What happens after that?

Shooting gives away the position of directly fired weapons. (We shall ignore for the moment weapons triggered by remote control or 'tripwires'.)[5] So, as the defenders fire more shots and reveal their positions, whilst the attackers' positions remain known as a result of their movement and firing, the position will tend to become *symmetrical*: both will tend to have an equal knowledge of the positions of the other. At this stage there seems no reason why there should be an advantage to defender or attacker, except in so far as the defender may have better protection, for example, through prepared earthworks and trenches, and will have better knowledge of the terrain. In other words, there are always the classical advantages enjoyed by the defender. But leaving these aside, there seems to be a phase of symmetry once the defender has revealed his position by firing.

At that stage, each side can move so as to seek concealment and protection or to disengage. But whatever the outcome of the local engagement, the attacker must advance again (until the day when his opponent decides to offer no more resistance). In advancing he will again reveal himself and make himself vulnerable to first shots by the defender – if the defender has adopted a dispersed defence in depth that takes advantage of the cover which, *ex hypothesi*, is available.

To sum up, as accuracy rises, it is increasingly important to be the first to shoot. The law that it pays to concentrate for the sake of numerical superiority ceases to be generally valid: it may pay to disperse, so as to be concealed and hence to be in a position to shoot first.

Whether the attacker or defender will gain from high accuracy (and high kill probabilities per shot) depends on the terrain and on the efficiency of target surveillance. Those two factors determine whether it is the attacker or defender who enjoys the advantage of shooting first. If the terrain offers little cover and/or the means of surveillance are efficient, the advantage goes to the one who chooses to shoot first, that is the attacker. Where the terrain offers good cover and/or the means of surveillance are poor, the advantage will go to the defender.

The oceans (as regards surface ships), airfields and the desert areas of the Middle East look like places where accuracy will help the attacker. Europe – though the extent of cover varies – looks like an area where accuracy will help the defender in land warfare. An important qualification is that the increasing importance of counter-measures is likely to increase the uncertainty of each side in a confrontation as to the potential performance of its own weapons and those of its opponent.[6]

Notes

1. British Atlantic Committee, *Diminishing the Nuclear Threat: NATO's Defence and New Technology* (London, 1984), pp. 30–3.
2. It is a proposition the author came across in the course of work under way with Anders Boserup on the theory of strategy. An early version will be found in 'Accuracy and Lanchester's Law: A Case for Dispersed Defence?', in P. Bennett (ed.), *Analysing Conflict and its Resolution: Some Mathematical Contributions* (Oxford, 1987).
3. See Michael Howard, 'Men Against Fire: Expectations of War in 1914', and Stephen Van Evera, 'The Cult of the Offensive and the Origins of the First World War', *International Security*, vol. 9 (1984–5), no. 1, pp. 41–57 and 58–107.
4. F. W. Lanchester, *Aircraft in Warfare: The Dawn of the Fourth Arm* (London, 1916), chapter 5.
5. Tripwires here means any trigger activated automatically by the approach of a person, vehicle or other weapon, for example mines, booby-traps or homing-devices activated by sensors.
6. This chapter is a shortened version of an article previously published in *Arms Control*, vol. 7, no. 3, May 1986. The author is grateful to the editor and publisher for permission to reproduce it here.

8 Mutual Defensive Superiority and the Problem of Mobility along an Extended Front
Anders Boserup

INTRODUCTION

Today's highly mobile and heavily armoured military formations are all-purpose forces, designed for manoeuvre warfare and for attack as well as defence. With such forces, equality between East and West (if it could be defined) would not guarantee security and military stability. Whatever one side deems sufficient for defence is bound to be regarded as excessive by the other side, as each can only buy security at the other's expense.

If the security concerns of the two sides are to be reconciled, the character of the forces must be changed so as to ensure that they are much stronger when fighting on the defensive than they are when fighting on the offensive. It is therefore not the equality or parity of the forces on the two sides but the inequality of offensive and defensive strength which is the condition for genuine balance, for equal and assured security and for any meaningful concept of sufficiency. This can be expressed in terms of the defensive (D) and offensive (O) capabilities of the two sides (a and b) as a stability requirement involving two inequalities, both of which must be satisfied:

$$\begin{cases} D_a > O_b \\ O_a < D_b \end{cases}$$

This is the condition of 'mutual defensive superiority'. It is the true condition of balance and stability at the military level and the only possible basis for common security.

In the past, it was mostly taken for granted that mutual defensive superiority would have to be achieved through unilateral steps. In this perspective, the first inequality expresses the need for a to maintain adequate defences, while the second expresses the goal of an overall military capability which cannot be perceived as a threat by the other side. To satisfy both inequalities, one is faced with the difficult task of simultaneously increasing the capability for defence, D_a, and decreasing the capability for attack, O_a. In this unilateralist perspective the key question was how to design military forces which are much more potent in defence than in offence. Not surprisingly, the debate focused almost exclusively on military issues, while analyses of the wider political benefits of mutual defence dominance seemed quite premature. Unless it could be shown beyond any possible doubt that non-offensive defence was sound and feasible, even in the narrowest military terms, it would clearly be a non-starter anyway. The discussions thus tended to focus on technical issues with little appeal beyond a narrow circle of experts and on the difficulties and risks, real and imagined, in a complete restructuring of military forces. Clearly, the fear that ill-considered experiments might jeopardise security deserve to be dealt with in a serious fashion. All the same, the debate has been distorted by the tendency to concentrate on narrow issues of military efficiency, while ignoring the positive implications of a security regime based on defence dominance for arms control and for undercutting tension and suspicion.

Recently, however, the conditions for debate and reflection on these issues have changed radically. In the Budapest Address of June 1986, the Warsaw Treaty members recognised the need to base the military concepts and doctrines of the two alliances on defensive principles. They proposed to work out procedures for the reduction of armed forces and armaments such that this process would lead to the lessening of the danger of a sudden attack and promote the consolidation of military–strategic stability on the European continent. In December of the same year, the member states of the North Atlantic Treaty also declared that future arms control should focus on the elimination of the capability for surprise attack or for the initiation of large-scale offensive action. Subsequent statements at the highest political levels have elaborated on these positions. As a result, it is now possible to think of the idea of defence dominance, not solely in terms of unilateral implementation, but also as a process in which East and West jointly work towards a regime of mutual defensive superiority as a basis for common security and for radical reductions in arms.

When the achievement of defence dominance is seen in this light entirely new aspects come to the fore. The political and military benefits become readily apparent and most of the former doubts and objections of a military nature lose all force. Instead of the difficult and costly task of building up sufficient defensive strength to match the opponent's present and projected offensive capability, one can imagine achieving the same result through joint reduction by both sides of the most threatening and destabilising force components. Above all, reductions need not come only at the far end of a long process of military restructuring but could be part of the process from the outset.

The prospect of a joint approach to mutual defensive superiority does, however, catch us somewhat unprepared. Much thinking has been devoted to designing failure-proof defence systems for unilateral or 'two-sided unilateral' implementation, little to ways of building down offensive capabilities by common accord. It would seem that many former ideas and concepts will have to be re-examined in the light of these new possibilities. This chapter is an attempt to reconsider some of the issues dealt with in the past in a perspective which brings out more clearly the relationship between unilateral and joint approaches to mutual defensive superiority.

Confusion between stability at the tactical level and at the level of the theatre of operations, and failure to distinguish attack, army against army, from offensive manoeuvres at a lower level, lie at the heart of much of the present obsession with the overall numerical balance of forces. This chapter accordingly tries to identify the determinants of stability at the operational level and to relate them to the determinants of stability at the tactical level.

SUPERIORITY OF THE DEFENCE

Like other types of forces, today's armoured divisions are inherently much more effective in defence than they are in attack. When they attack they are exposed and visible and must rely on complex and vulnerable support and logisitics. Meanwhile, the defending forces can remain under cover, evade the initial blow, interpenetrate the attacking columns and wait for the best time and place to strike back. It is generally assumed that in a well-prepared defensive combat, all-purpose forces like those presently deployed in Europe can hold back attacking forces numerically superior by a factor of three. To be reasonably certain of breaking through the defences, an attacker would need a superiority significantly better than this, say of the order of 5:1.

Such figures are, of course, only crude rules of thumb, if not completely misleading. They are actually derived from the experience of the First World War and may not adequately represent present conditions. Basil Liddell Hart noted that Allied attacks in the Second World War rarely succeeded unless the attacking troops had a superiority in strength of *more than* 5:1, *together with* domination in the air. As a striking case he cited the attempted Allied break-out from Caumont in the Normandy campaign. The Allies had succeeded in concentrating and launching

> two specially strong army corps against a ten-mile sector held by only two weak German infantry regiments. The attacker's superiority in fighting units was nearly *10 to 1*, and in number of troops was more than that. Being backed by air supremacy, the real measure of advantage must be reckoned as at least 20 to 1, and may well be reckoned as 30 to 1. Moreover, a total of well over 1,000 tanks were concentrated, in this case, on a sector where there were no German tanks in the earlier phase of the battle. Yet the massive blow failed to overcome the thin defence except on the western part of the sector, and even there it was checked on the third day when meagre tank reinforcements began to arrive on the German side. And it suffered continuous checks during the days that followed.[1]

In spite of this it will be assumed that fixed ratios for successful defence and attack can be defined. The number of attacking units that one defending unit can safely hold back will be referred to as the 'hold-factor', H. It will mostly be assumed that H is around 3. Similarly, the superiority required to be reasonably certain of breaking through a front line will be referred to as the 'break-factor', B. This factor is assumed to be around 5. The combat advantage of the defence, the 'defence:offence ratio', is measured by the *product* of the two numbers H and B: if five units of force can fend off fifteen units when fighting in the defensive but cannot overwhelm more than one unit when attacking, then these units are *fifteen* times more effective in defence than they are in attack. In combat, therefore, defence enjoys a vast superiority of one, possibly two orders of magnitude, even with today's armoured forces which were actually optimised for offensive action.

At the level of the battlefield, therefore, mutual defensive superiority is easy to achieve, unless the disparity of force is very large indeed. But this stability at the level of the battlefield does not translate into overall stability at the operational level.

CONCENTRATION AND SURPRISE

The attacker has only one significant advantage over the defender: that of choosing time and space through concentration and surprise. Once the effect of surprise wears off, all the advantage is on the side of the defender. He can fight back from prepared positions and need not expose himself to enemy fire, while the attacker must bring forward his troops, equipment and supplies. It is these advantages for the defence which are summarised in the defence:offence ratio.

The attacker's advantage of concentration and surprise must be considered at the level of the entire theatre of operations. Figure 8.1 illustrates the situation for an extended front. As shown, the superiority of the defence allows the attacker to release a part of his forces all along the front and bring them together over a shorter sector of the front, building up sufficient strength in this sector for a breakthrough. In the other sectors of the front, however, he will have to leave enough forces behind to guard against the possibility of counter-attack, that is, enough forces to ensure that his opponent cannot develop a superiority in excess of the 'hold-factor'.

The basic point is very simple. To overcome the defence and break through along one sector of the front, the attacker must build up a superiority in that sector better than B:1, where B is the 'break-factor' as previously defined. This must be done by bringing in forces from other sectors of the front, taking care, however, not to deplete these sectors to such an extent that he becomes vulnerable to a counter-attack which might jeopardise the entire plan of operations. The attacker must therefore leave enough forces behind to ensure that the opponent cannot achieve a superiority better than H:1, where H is the 'hold-factor'. If the 'break-' and 'hold-factors' are known, as well as the ratio, S, between the total combat strength of the attacker and of the defender (A and D respectively) it is easy to calculate how many units can be released for the spearhead. This again determines the maximum width of the spearhead. This maximum width, called the 'assault width', will be used here as a simple measure of attack capability. The 'assault width' is thus defined as that proportion of the front over which the presumed attacker could achieve a superiority of B:1 while maintaining an inferiority no worse than 1:H along the remainder of the front. The assault width is given in kilometres, it being assumed throughout that the front has a total length of 800 kilometres, the distance from the Baltic to the Alps. Attack capability could have been measured in many other ways, for

H: 'Hold-factor'
   (number of attacking units which one unit can hold back)
B: 'Break-factor'
   (number of attacking units needed to break through one unit)
N: NATO strength per km.,   *W*: Warsaw Pact strength per km.

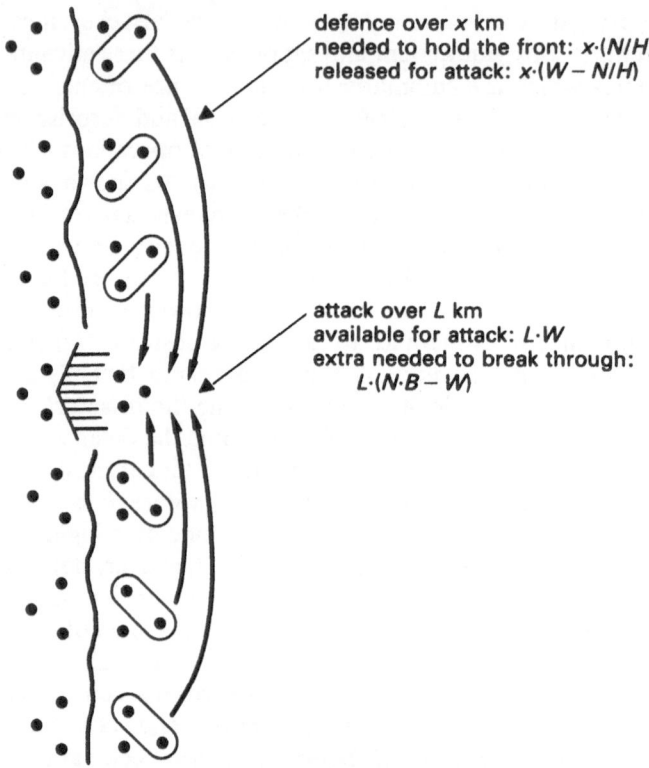

defence over *x* km
needed to hold the front: $x \cdot (N/H)$
released for attack: $x \cdot (W - N/H)$

attack over *L* km
available for attack: $L \cdot W$
extra needed to break through:
   $L \cdot (N \cdot B - W)$

A breakthrough is possible over *L* km of an 800 km front when
   $$L \cdot (N \cdot B - W) = (800 - L) \cdot (W - N/H)$$

hence:

$$L = 800 \cdot \frac{S \cdot H - 1}{B \cdot H - 1}$$

where *S* is the strength ratio: attacker/defender
($S = W/N$ or $S = N/W$, as the case may be).

Example: if $H = 3$ and $B = 5$ and the Warsaw Pact: NATO force ratio is 3:2, then Warsaw Pact forces could break through over 200 km of front, while NATO (having a 2:3 force ratio) could break through over 57 km of front.

Figure 8.1   Concentration for attack

example, by the number of fresh units the attacker has left for operations in the enemy rear after breaking through over a sector of some specified length (20 kilometres for example). But this would not significantly affect the conclusions of the present analysis.

Figure 8.1 explains the basic model in greater detail and shows that the assault width, *L*, is given by the expression:

$$L = 800 \frac{S \cdot H - 1}{B \cdot H - 1}, \text{ where } S = A/D$$

There is mutual defensive superiority if the assault widths are negative for both sides. It is evident that this cannot happen with the assumptions thus far built into the model. One side's assault width will only be negative (meaning that it cannot break through, even in a very small sector) if that side is outnumbered by more than a factor *H* (so that $S \cdot H$ is less than 1 in the above equation). If it is, the other side cannot be outnumbered as well.

In reality, an attack must, of course, develop over a certain length of the front, say, 20 kilometres or so, and one could regard an arrangement in which both assault widths are less than this as essentially stable. It will be shown that even this limited form of stability is not easily achieved along an extended front. Assuming, as an illustration, that the ratio of strength between Warsaw Pact and NATO forces were 3:2, then the assault width of the Warsaw Pact (*S* = 3:2) would be 200 kilometres and that of NATO (*S* = 2:3) 57 kilometres. Even with a 2:3 inferiority the weaker side has ample means for attack.

It is clearly not for lack of a favourable defence:offence ratio that mutual defensive superiority cannot be established with today's all-purpose forces. Figure 8.2 shows how the assault widths are affected by the 'hold-factor'. Even if this factor were much larger than 3 the situation would be essentially the same. Above all, it should be noted that it is not only the stronger side which is able to concentrate enough forces for a breakthrough. So is the weaker side. To foreclose this option the stronger side would have to enjoy an *overall* superiority better than 3:1. The East–West force ratio (in real combat strength, rather than raw numbers) is certainly nowhere near this figure. It is often claimed in the West that the Warsaw Pact could mount a successful conventional attack by a suitable massing of forces, whereas NATO simply lacks the forces to do so. As it stands, this claim is clearly wrong and results from an inadequate appreciation of the extent of the instability inherent in present force

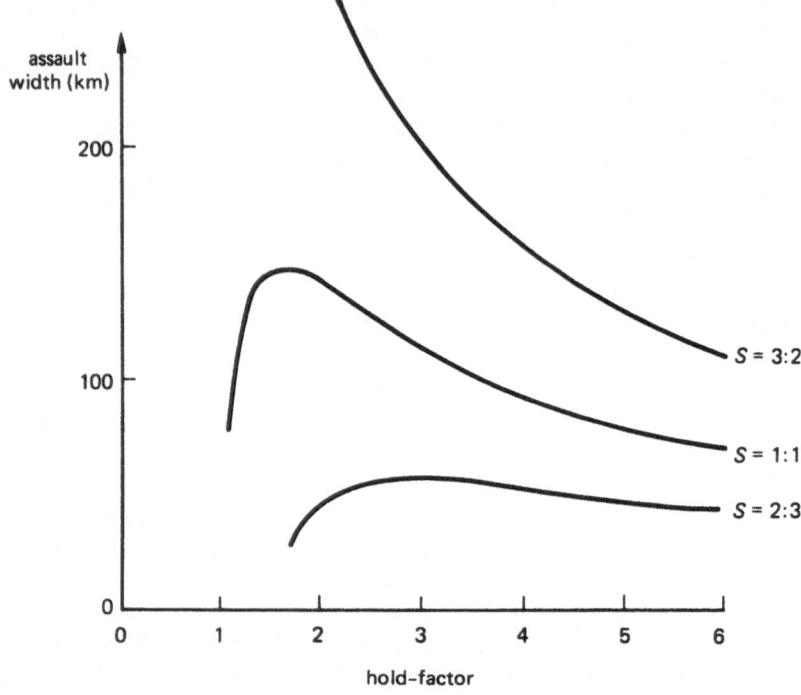

Attack capability is shown as a function of H (assuming $B = (5/3) \times H$), for attack: defence ratios of 3:2, l:l and 2:3 respectively.

Figure 8.2   Influence of the 'hold-' and 'break-factors' on attack capabilities (mobile forces only)

structures. It is the mobility of the present forces, rather than the numerical disparity between East and West, which is the main source of military instability.

DEFENSIVE BARRIERS

To implement a non-offensive defence posture on a unilateral basis, the most straightforward idea is to change gradually the military structure from mobile all-purpose forces to stationary forces which can fight effectively in a defensive mode, but which would be useless for attack because they cannot move or cannot move forward.

Several ideas of this kind have been proposed. One is the fire-belt where dug-in sensors along the front would reveal the location of attacking units and guide the fire from mine-throwers and artillery situated in the rear. Another is the net zone where small detachments of light infantry, dispersed and concealed in prepared positions, would confront an intruder with various kinds of precision-fire and supply target information to more distant artillery and rocket units. All zones or combinations of zones of this kind are referred to here as 'defensive barriers'.

Defensive barriers will be characterised by their 'efficiency', e, defined as the proportion of the attacking forces which would be destroyed by the barrier. Efficiency thus defined runs from zero (when the barrier is non-existent or completely ineffective), to one (when the barrier is impenetrable).

To determine the effect of a barrier let F be the number of attacking units engaged at the spearhead. If the efficiency of the defender's barrier is e it will absorb eF attacking units, and the mobile forces of the defender must therefore cope only with the remaining $(1-e)F$ enemy units. It is as if the 'break-factor' had been increased from B to $B/(1-e)$.

Along the remainder of the front the situation is similar. To foreclose counter-attack the attacker, when there is no barrier, must leave behind $1/H$ units for each unit of the defender. When the attacker is defended by a barrier of efficiency e', a counter-attack would be depleted by a factor e'. For every defending unit the attacker must therefore leave behind $(1-e')/H$ units. It is as if the 'hold-factor' had been increased from H to $H/(1-e')$. The formula for the assault width is therefore analogous to the previous one:

$$L = 800 \frac{S \cdot H/((1-e') - 1}{B \cdot H/(1-e)(1-e') - 1}$$

where e and e' are the efficiencies of the barriers of the defender and the attacker respectively.

The curves in Figure 8.3 show how attack capabilities (measured as always by assault widths) change, as barrier efficiency increases from 0 to 1. The three upper curves describe the situation when the stronger side (with a strength ratio of 3:2) attacks and there are barriers on both sides (the fully drawn line), only on the side of the attacker (the upward sloping broken line), or only on the side of the defender (the downward sloping broken line). In the same way, the three curves at the bottom show what happens if the weaker side attacks (strength ratio of 2:3).

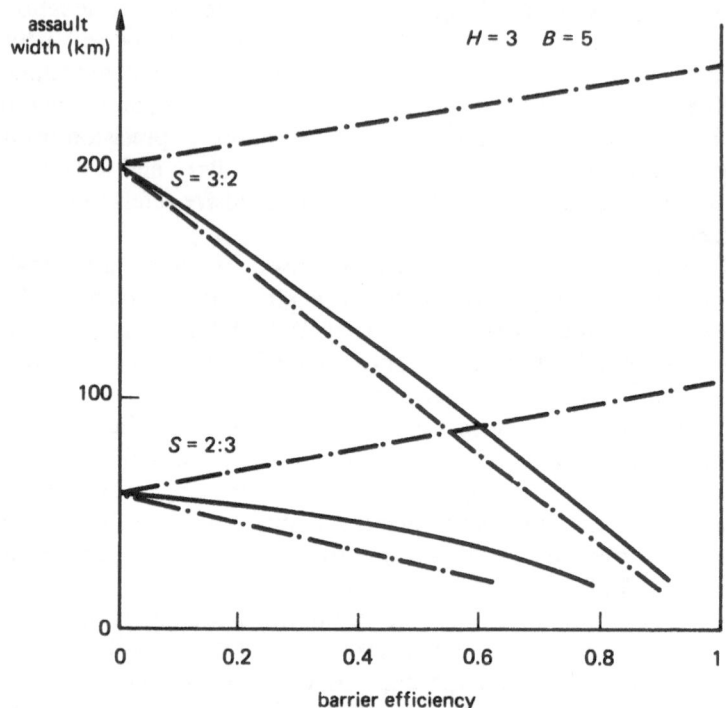

Attack capability shown as a function of barrier
efficiency. Top: $A:D = 3:2$. Bottom: $A:D = 2:3$.

Fully drawn curves; equal barriers on both sides
Upper dotted lines: barrier on defender's side only
Lower dotted lines: barrier on attacker's side only

Figure 8.3   Effect of adding barriers on one side or on both

The most important point to note is that such defensive barriers
have surprisingly little effect on overall military stability. True
stability (assault widths of zero) simply cannot be achieved, except in
the trivial but unrealistic case of completely impenetrable barriers on
both sides. For barrier efficiencies which might be achieved in
practice, the contribution to stability is not very impressive. As the
upward sloping curves show, a barrier actually increases the attack
capability if it is not matched by a similar barrier on the other side.
The reason is that the more effective the defence along one sector of
the front, the more forces can be released for attacking in other
sectors. This underlines the point so often made that it is nonsense to

describe certain weapon systems or force components as being by nature 'defensive'. Even the barrier forces which, by assumption, cannot possibly take part in an attack may contribute to the overall attack capability. Whether a particular force component will enhance defensiveness and military stability or will undermine it can only be determined by looking at its place in the overall arrangement of forces on both sides, as is being done here. In summary, Figure 8.3 shows that a unilateral defensive barrier does not promote mutual defensive superiority. It enhances one's own capability for defence, but instead of diminishing the threat to the potential opponent, it actually increases it.

Defensive barriers are more promising in the context of joint attempts to achieve mutual defensive superiority. At least the effects are unambiguously positive. There is a reduction of attack capabilities on both sides, and, as shown by the convergence of the fully drawn lines in Figure 8.3, the barriers tend to attenuate the initial disparity of force. All the same, the contribution to stability is fairly modest and the result is a far cry from true mutual defensive superiority. To build barriers which can absorb, say, 30 per cent of an attack anywhere along the dividing line in Central Europe would clearly be a costly and difficult matter. And yet it would only reduce attack widths from 200 to 140 kilometres and from 57 to 49 kilometres respectively.

DEFENSIVE BARRIERS WITH FORCE REDUCTIONS

It might be thought that mutual defensive superiority could be achieved if the establishment of defensive barriers were accompanied by suitable reductions in the all-purpose forces, since it is the latter which caused military instability in the first place. In fact, this is not so.

Referring first to the case of unilateralism, let it again be assumed that the NATO:Warsaw Pact force ratio is 2:3. Figure 8.4 further assumes that a defensive barrier of efficiency 1/3 has been erected on the Western side and none on the Eastern side. The curves show how reductions in the mobile forces would affect the assault widths. The barrier itself, in the absence of any force cuts, brings down the Warsaw Pact assault width from 200 to 130 kilometres and raises NATO's assault width from 57 to 74 kilometres as could already been seen from Figure 8.3. As the forces on the Western side are reduced, the assault width for the Warsaw Pact rises again (upper curve), while

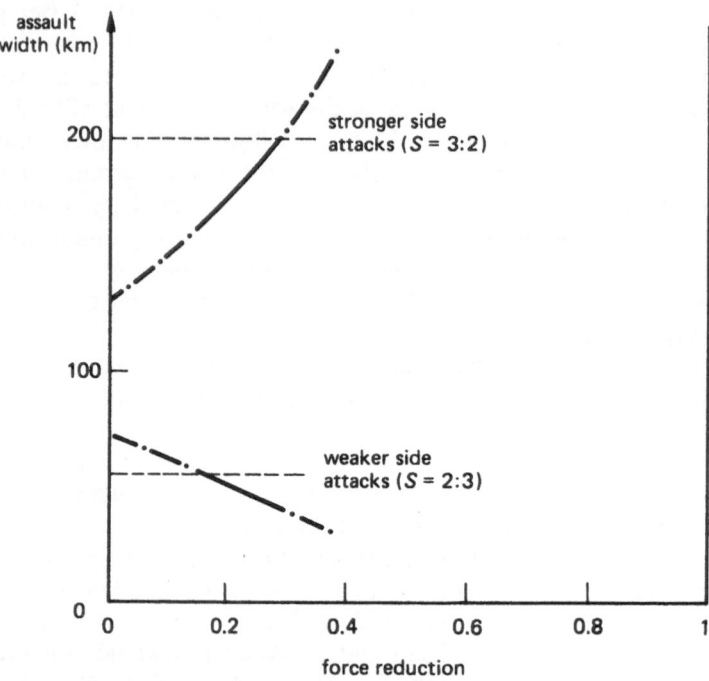

Attack capability shown as a function of the force reduction when the barrier efficiency is 1/3.

With the stated assumptions a 15–30 per cent force reduction would improve defensive capabilities on both sides.

Figure 8.4   Barrier and force reduction on the weaker side only

the assault width for NATO declines (lower curve). If the combination of barrier and force cuts is to leave Western security undiminished, the force cuts may not exceed 30 per cent as shown by the upper curve. If it is to leave the security of the Warsaw Pact undiminished the force cuts must be at least 15 per cent as can be seen from the lower curve. Figure 8.5 shows a similar situation for a combination of defensive barrier and force cuts on the Eastern side. In this case, the force cuts must be in the range from 5 to 20 per cent if the aggregate effect is to improve the security on both sides. However, the two figures also show that only modest improvements in security can be achieved in this way: a reduction of assault widths of the order of 20 kilometres, 40 at most, on one side or the other.

It remains to see what happens if barriers and force cuts are

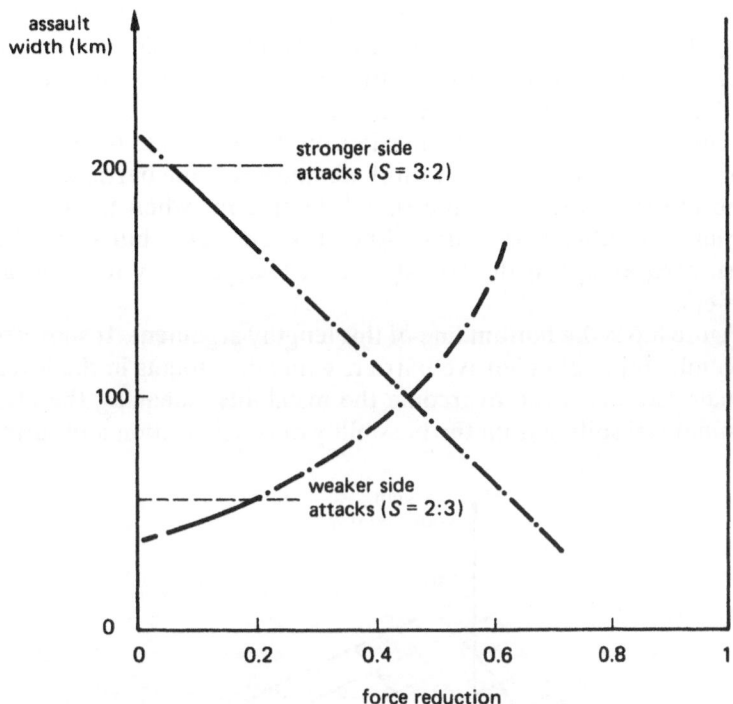

Attack capabilities shown as a function of the force reduction when the barrier efficiency is 1/3.

With the stated assumptions a 5–20 per cent force reduction would improve defensive capabilities on both sides.

Figure 8.5 Barrier and force reduction on the stronger side only

introduced on both sides in a joint attempt to institute a regime of mutual defensive superiority. In this instance, there are so many different parameters that the whole story cannot be told in one simple diagram. To simplify matters, therefore, the issue has been reduced to its essentials by asking what is *the very best* that can be achieved if the two sides are actively cooperating to reduce attack capabilities to a minimum and are adjusting the levels of their all-purpose forces accordingly. It is being assumed, therefore, that both sides are trying to minimise not only the assault width of the other side but also their own, so as to eliminate threat perceptions and achieve mutual defensive superiority. Even with this extreme assumption about cooperative behaviour and complete lucidity, attack capabilities cannot be reduced to zero. A residual remains, and all attempts to bring the attack capability of one side down below this minimum will push the attack capability of the other side up above it. This irreducible minimum is shown in Figure 8.6 as a function of the effectiveness, *e* and *e'*, of the barriers on the two sides. As one would expect, this minimum assault width diminishes as the barriers become more efficient. From a value of 115 kilometres when there are no barriers on either side it diminishes towards zero, but it does not reach zero except in the trivial case of two perfectly impenetrable barriers.

Figure 8.6 is the bottom line of this lengthy argument. It shows that no combination of defensive barriers with adjustments in the level of all-purpose forces can overcome the instability (meaning the attack capability) resulting from the possibility of concentration and surprise

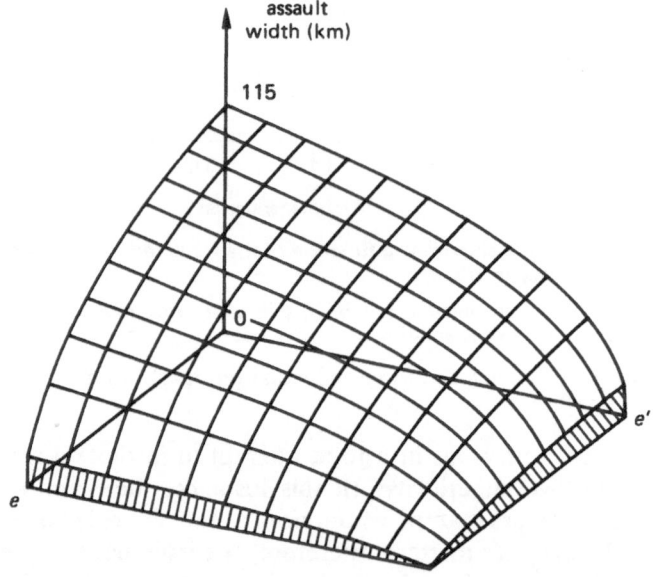

Minimum assault widths are shown as a function of the barrier efficiencies *e* and *e'*.

For each pair (*e*, *e'*) force reductions were adjusted to minimise the largest of the two assault widths.

(Only the lowest possible assault widths are shown)

Figure 8.6   Barrier and force reduction on both sides

along an extended front. Defensive barriers can diminish this instability, and appropriate adjustments in force levels may further diminish it, but it cannot be eliminated. As Figure 8.6 shows, it cannot even be brought down to an insignificant level. Unless the defensive barriers are virtually perfect, one side, the other, or both, will retain the capability, theoretically, for breaking through over a substantial part of the front.

DEFENCE VERSUS DELAY

In the past, non-offensive defence has been thought of mostly in terms of unilateral action. While such unilateral action might later be followed up by the other side, the first task was to convince doubters in one's own camp that military options could be developed which would be ill-suited for attack and yet as effective, if not more effective than present military postures in defensive combat. For this reason, it seems, it has mostly been taken for granted that the purpose of a defensive barrier was to provide effective defence, that is, to wear down the forces that might try to break through.

The conclusion of the preceding analysis would seem to be that this idea is flawed. At least in the case of an extended front, defensive barriers cannot create a situation of mutual defensive superiority. Along an extended front, what is needed to give overwhelming superiority to the defence is not so much a defensive barrier which can impose a high attrition rate on the attacker; it is rather a delaying barrier which gives to the defence the time needed to redeploy its forces. Indeed, it has been shown that present-day, all-purpose forces are actually very weak in attack against similar forces if the latter are well prepared and dug in. Only through concentration and surprise can an attacker hope to break through, and he can do so only if he can retain the advantage of surprise and can bind the forces of the defender in a pattern which is quite irrational from his point of view.

A delaying barrier would bring the superiority of the defence fully into play. Every unit which can be brought in from another sector of the front early enough to be able to fight defensively at the point where the attacker tries to break through, compels the attacker to bring in another five units. Every unit which arrives too late and which must fight back is less effective by at least an order of magnitude. The principal merit of a delaying barrier is to free the all-purpose forces from guard duty along the front and transform them into the equivalent of mobile reserves. By eliminating the possibility of offensive concentration and surprise it brings out the stabilising potential inherent in the 15:1 defence:offence ratio of these forces. The resulting mobile reserves, therefore, become a powerful stabilising element, provided they are combined with delaying barriers on both sides. This is so, even if they have the training and support which would be required for an offensive, for the point is that each of these reserve units compels the other side to find five additional units if he plans to attack but only one-third of a unit if he is only contemplating defence.

This again illustrates the fact that force components, like weapons, are not 'offensive' or 'defensive' *per se*. Alone, as they are now, or in combination with an attrition barrier, today's all-purpose forces are a factor of instability. Combined with a delaying barrier they would be a stabilising factor.[2]

Notes

1.   Basil Liddell Hart, *Deterrent or Defence* (London, 1960), p. 179.
2.   This chapter was submitted to the Pugwash Study Group on Conventional Forces in Europe at its Sixth Workshop held at Altamura, Italy, from 1 to 4 October 1987.

9 Military Stability and Defence Dominance at the Tactical Level

Anders Boserup

INTRODUCTION

This chapter is part of a wider study of the concept of conventional stability. Here a number of simple analytical approaches, concepts and models are developed to bring out the relevant aspects of the concept, to examine the relationships between stability and defence dominance, and to determine how these are related to specific military structures and doctrines.

Strategic analysis is essentially a calculus of ends and means, and lends itself to a subdivision with a hierarchy of distinct analytical levels: *strategy* (military objectives); *operations* (theatre and movement); *tactics* (close combat); and *force design* (hardware and organisation). At each level the question of military stability assumes a different form, depending on the specific problems and mechanisms, and distinct models are therefore needed at each separate level. Table 9.1 summarises this hierarchy of levels. Starting from the top, there is a specific set of objectives at each level and specific approaches to meet them. Such 'approaches' are in effect models. They call for certain means, and describe how these means act together to bring about the desired objectives within the general context provided by the approaches adopted at higher levels. These means then determine the objectives for the next level below. Needless to say, the elaboration of approaches at the higher levels must constantly be revised in the light of the material constraints which limit the free choice of means at the lower levels.

At the level of strategy the basic concept is that of Mutual Defensive Sufficiency. This has been discussed in many papers, and it has been shown that at the level of strategy this is essentially equivalent with the idea of military stability.[1] At the operational level the key issue is the attacker's advantages of concentration and surprise. A model designed to examine this question in the case of a forward defence along an extended front has been developed else-

Table 9.1   Levels in means-ends calculus

Level	Objective	Approach	Type of model
Policy goals	Remove fear and unwind the arms race	Non-threatening posture	
Strategy (military aims)	Military stability (attack unattractive)	Superiority in defensive combat mode	Mutual defensive sufficiency
Operations (war plans)	Counteract surprise and concentration	Attrition and delay forces	Theatre mobility and barriers
Tactical (engagement)	Fend off or wear down massed attack	Concealment and dispersion	Combat dynamics (Lanchester)
Force design	Maximise the defender's exchange ratio	Early detection and accurate indirect fire	Geometry of weapons interactions

where in this volume.[2] Subsequent studies have extended this model in several directions.[3] As these models suggest, military stability at the operational level is best achieved through a combination of stationary, barrier-type forces and mobile armoured formations. The purpose of this chapter is to develop models to determine how such forces would interact in combat, that is, at the tactical level.

Combat dynamics are best studied by means of Lanchester-type models which use differential equations to describe the attrition of the forces of the two sides as combat develops and casualties accumulate.[4] A variety of models (such as quadratic and linear) can be constructed in this way, depending on the precise form of these equations. The main interest here is in asymmetric models in which the two sides differ in their mode of combat.[5] Such a model can reflect the distinction between offence and defence in its mathematical structure, and it can therefore be used to analyse the conditions for stability and defence dominance (at the tactical level, of course).[6]

THE BASIC, QUADRATIC LANCHESTER MODEL

Lanchester's quadratic model describes the dynamics of a battle between two belligerents, here called Red and Blue. The forces on each side consist of 'combat units'. These could be tanks, aircraft, infantrymen, ships or any other entity which can find enemy combat units and destroy them. The model can also describe asymmetric combat, for example, an engagement between armoured vehicles and anti-tank weapons, or between ground-attack aircraft and surface-to-air missiles. The basic model will be examined in some detail as a background to the modifications which will be introduced later.

The quadratic model is based on the following assumptions:

● all Red combat units are of the same kind, and so are all Blue combat units;

● combat units are either fully operational ('live') or completely disabled ('killed');

● combat units can distinguish live targets from disabled units, and they only fire at live units;

● live combat units are always active, that is, they go on engaging enemy units until they are themselves disabled;

● a live combat unit kills enemy units at a constant rate, the 'kill-rate', which is independent of the number of enemy units.

The first three of these assumptions are introduced only for the sake of simplicity. If the object were to reproduce actual combat conditions as accurately as possible they could easily be relaxed to allow for more than one type of combat unit on each side, for partially damaged units and for fire wasted on targets which are already dead. Here, however, the aim is to examine the overall analytical properties of the model, and this is best done by keeping it as simple as possible.

The critical assumptions lie fourth and fifth. They determine the dynamics of combat and, with it, the place and function of combat in overall operational plans. At the same time, their validity depends on force design (the characteristics of weapons and detectors in terms of effectiveness, ranges and disposition on the battlefield). These two assumptions (and their alternatives) therefore establish the link between models at the tactical level and models at the levels above and below. This will become clearer as different models of combat dynamics are examined.

On the basis of these five assumptions, the development of combat can be expressed as a pair of first-order differential equations which describe how Blue casualties are generated by Red fire, and vice versa. Let $R(t)$ and $B(t)$ denote the number of Red and Blue units surviving at time t, and let $R(0)$ and $B(0)$ denote the number of units at time zero when combat begins. The decline in the number of

Blue units per unit of time, $-dB(t)/dt$, is then given by the number of Red units $R(t)$ times their kill-rate, k_R. A similar equation describes the decline in the number of Red units, $-dR(t)/dt$, in terms of the number, $B(t)$, and kill-rate, k_B of the Blue units.

> *Basic, quadratic model*
> Blue losses: $-dB(t)/dt = k_R.R(t)$
> Red losses: $-dR(t)/dt = k_B.B(t)$

To solve these equations is to determine the two functions, $B(t)$ and $R(t)$, which describe how the numbers of live Blue and Red units develop over time. The precise shapes of these functions depend on the values assigned to the kill rates and on the number of combat units at the onset of battle. An example is shown in Figure 9.1. In this particular case Red has an initial numerical superiority. It grows in the course of combat until, at time t_F, all Blue forces have been wiped out and the surviving Red forces, R_S, have sole control of the battlefield.

MILITARY CAPABILITY

There is no such thing as military capability *per se*. The combat potential of a given system of arms and armed forces will obviously depend on the mode of combat. The different items in an inventory combine in one way, with one set of weights, when employed, say, in defensive combat, in mountainous terrain and according to some specific tactical scheme. In other forms of combat, with other types of tactics, the same items combine in another way with another set of relative weights. It is evident that the peacetime inventories of armaments and armed forces commonly presented as evidence of the 'balance' or 'imbalance' in the military capabilities of East and West are completely meaningless as they stand, if only because any assessment of military capability must relate to performance in combat and must therefore start with a specification of the mode of combat being considered.

In the basic, quadratic Lanchester model (and in other models to be considered below), the character and size of the forces on the two sides, as well as their mode of interaction in combat, have been fully specified. In these cases, military capability can therefore be defined in a strict and meaningful way. The key point about a meaningful measure of military capability is, of course, that it must rate as equal

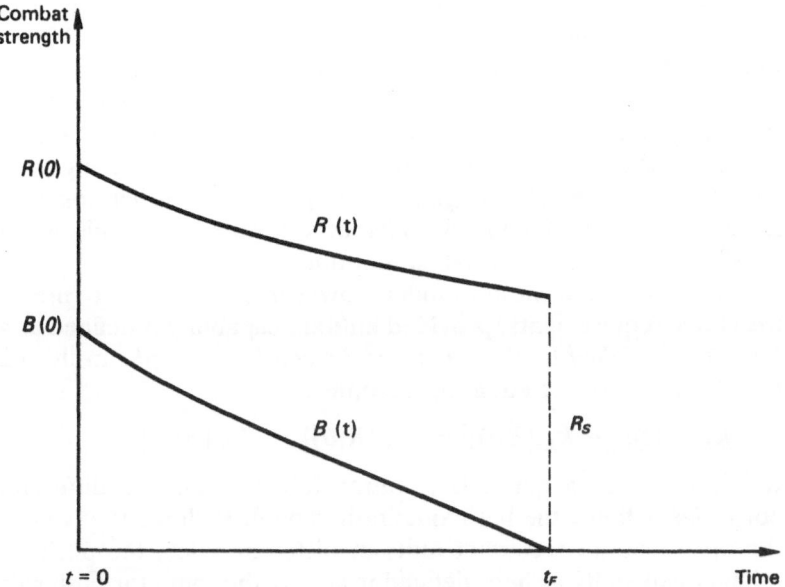

Figure 9.1   Decline in Red and Blue units during combat

forces which would precisely neutralise or annihilate each other in actual combat. In other words, the measure must be such that one can describe combat simply as *the mutual annihilation of military capability in a 1:1 ratio*. Denoting $C_i(t)$ the military capability of country or alliance i at time t, the military capabilities of countries or alliances X and Y must then satisfy the relation

$$C_X(t) - C_Y(t) = C_X(0) - C_Y(0)$$

This is the fundamental relation which defines the concept of military capability without reference to any specific model of combat dynamics. It simply expresses the requirement that *the difference in military capability between the two sides must be invariant under combat*. Like a temperature scale, a scale of military capability defined on the basis of this relation involves an arbitrary zero-point and an arbitrary unit of measurement.

With a definition satisfying this equation, equality in military capability implies *military equivalence*: the forces perform equally well in actual combat and neutralise one another. When forces are not equal, *military superiority* can be defined as the difference between the military capabilities of the two sides. It remains constant throughout combat. Consequently, the military superiority of one side over the other at the outset, equals the military capability of those of his forces which would survive, after all the opponent's forces have been wiped out. Ideally, military capability should be an *intrinsic property* of the total armed forces of a country or alliance, that is, it should be possible to define it without referring to the character and combat strength of the opponent. For heterogeneous forces consisting of combat units of different types this is only possible under rather special assumptions.

For the basic Lanchester model, however, it is possible to meet all the above requirements, provided military capability is defined as *the kill rate multiplied by the square of the combat strength*. In this case the above invariance equation becomes:

$$k_R.[R(t)]^2 - k_B.[B(t)]^2 = k_R.[R(0)]^2 - k_B.[B(0)]^2$$

It is easy to show that this equation follows from the differential equations defining the basic quadratic model. It shows that whereas the opponents lose combat units at different rates, the decline in military capability as here defined is always the same for both sides. The fact that military capability varies as the square of the combat strength is *Lanchester's Square Law*.

According to this model (and to the other models below), combat ends with the complete annihilation of all the forces on one side, while the other side comes out with some of his forces unscathed. Apart from the fact that battles are rarely, if ever, fought to the finish, this seems to imply that there is a 'winner'. In fact, the concepts of 'winner' and 'victory' are entirely out of place. These models deal with a limited engagement, and 'victory', therefore, can (at best) mean tactical success. It cannot be assumed that there is any simple relationship between tactical success (prevailing in battle) and victory in the sense of strategy (achieving the ends of war). What tactical success in one battle means for the outcome of the war as a whole is a matter which can only be decided at the operational and strategic levels of analysis.

It is often claimed that because imponderables play such an important role in war, it is impossible to calculate anything, and that models of combat dynamics and measures of military capability like those introduced here are useless because they imply a naïve, deterministic view of war. Force comparisons and numerical disparities are, however, invariably portrayed as matters of great significance, and this is a tacit admission that war is susceptible to mathematical analysis, and that numbers matter for the outcome. Clearly, an account of military capability based on an analysis of combat dynamics is no more deterministic than one based on the usual 'counting of the pieces'. In fact, it is simply illogical to attach significance to the size of military inventories, while dismissing the mathematical analysis of their expected performance in actual combat.

REPRESENTATION IN PHASE-SPACE

The dynamic aspects of the model, particularly the question of stability and instability, are best studied in terms of a so-called phase-space diagram, a two-dimensional plot in which the number of Red combat units is shown along one axis, and the number of Blue combat units along the other axis. In such a diagram the state of the system at any given moment is represented by a point whose coordinates, $(R(t),B(t))$, are the numbers of surviving Red and Blue combat units at that moment. As combat progresses and casualties accumulate on both sides, the point moves left and down, describing a trajectory which begins at the point $(R(0),B(0))$, and which ends when it reaches one of the axes. At that point, if not before, combat must stop, as all the units on one side have been eradicated. The number of surviving units on the 'winning' side is given by the intercept along the axis.

There is one trajectory through every point in the plane, and all the trajectories taken together constitute a family of curves as shown in Figure 9.2. The general shape of these curves determines the dynamic properties of the underlying model. For Lanchester's quadratic model these phase-space trajectories are hyperbolae. They are given by the equation

$$k_R.[R(t)]^2 - k_B.[B(t)]^2 = S_{R/B}$$

where $S_{R/B}$ is a constant. Each curve corresponds to a different value of this constant, which is simply the military superiority of Red over Blue for that particular trajectory, as defined above: $S_{R/B} = k_R.[R(0)]^2 - k_B.[B(0)]^2$.

Referring to Figure 9.2, the main diagonal is the trajectory for combat when the military capability is the same on both sides. In such cases, combat to the finish leads to the total eradication of the forces on both sides, as the trajectory reaches the point of origin, $(0,0)$.

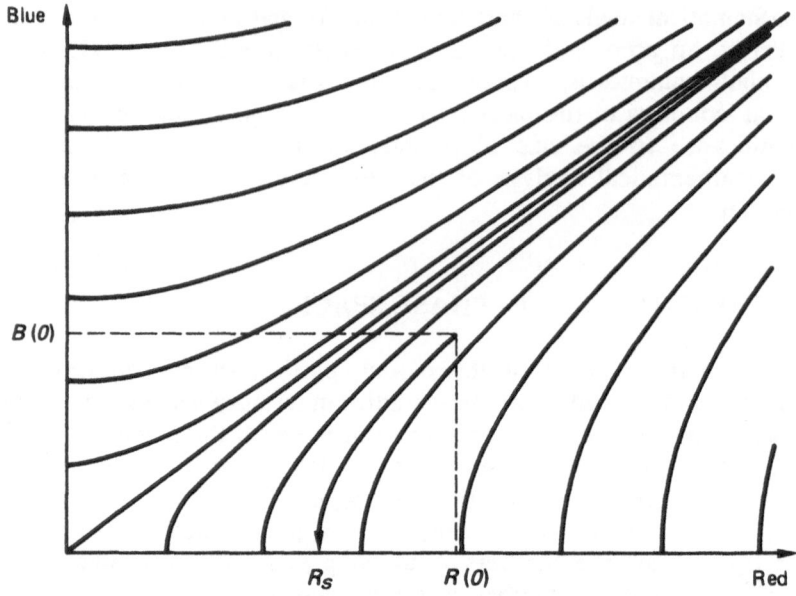

Figure 9.2 Phase-space trajectories for Lanchester's quadratic model

The diagonal divides phase-space into two parts. Points below the diagonal correspond to Red superiority ($S_{R/B}>0$). In these cases (of which an example is shown), combat follows a hyperbolic trajectory, ending on the Red axis when Red has R_s combat units left and Blue has none. In the same way, points above the diagonal correspond to Blue superiority.

Another important aspect of combat which can be directly read off from the phase-space diagram is the exchange ratio, which is defined as the number of enemy combat units killed per combat unit lost. This exchange ratio measures the average military 'effectiveness' of combat units, and is not to be confused with military capability as previously defined. The value of the exchange ratio at some particular point in the course of combat is simply the slope of the trajectory in phase-space at that particular point on the curve. Similarly, the average exchange ratio from start to finish is simply the slope of the straight line from the starting point, $(R(0),B(0))$, to the end point, $(R_s,0)$. For the quadratic model (but not for some models to be considered later) the exchange ratio is not only a function of the quality of men and weapons relative to those of the opponent, but it also depends on the ratio of friendly to enemy combat units: the more favourable the force ratio, the more favourable the exchange ratio.

BATTLEFIELD INSTABILITY

From the phase-space representation it can be readily seen that the combat situations described by the basic model are characterised by extreme instability. As shown in Figure 9.3a, slight differences in the ratio of forces at the outset are magnified in the course of combat, and have major implications for the final outcome. As already noted by Lanchester, this instability is particularly pronounced in the neighbourhood of the main diagonal, where any advantage secured by either side will tend to augment, and small initial disparities decide between major victory and major defeat.[7] This means that it is impossible to achieve two-sided military stability at the tactical level. Equality in military capability does not imply 'stability', it implies instability, for however carefully the military capabilities of the two sides are matched along each sector of the front, small last-minute reinforcements in one sector with forces from a neighbouring sector will upset the 'balance' and open the way for a successful attack.

Instability is further aggravated by the premium on surprise. Generally speaking, that side which opens the battle can expect to destroy a certain number of enemy units before actual combat begins. This is illustrated in Figure 9.3b. A straight combat would follow the normal hyperbolic trajectory from the point P_o, leading to a Red victory with a number of survivors equal to R_s. If Red attacks by

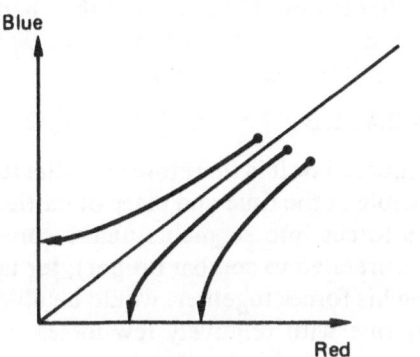

Figure 9.3a   Instability

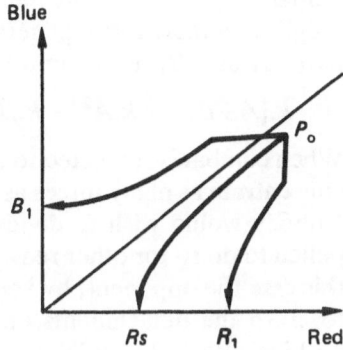

Figure 9.3b   Surprise attack

surprise, however, combat follows a trajectory which begins with a vertical drop (in this phase only Blue suffers casualties) and then continues along the relevant hyperbola. This trajectory ends with a more attractive result for Red, R_1. If Blue strikes by surprise the trajectory first moves left, and then follows a hyperbolic path with an outcome which is less favourable for Red, and which may even lead to victory for Blue at the point B_1.

At the tactical level, therefore, the situation is highly unstable as both sides have a military incentive to pre-empt. At the strategic level there can be a kind of military stability, provided one side has a clearcut overall military superiority, for in this case one side *need not* attack, and the other side *cannot* without losing. But if the two sides are roughly equal in military capability, pre-emption becomes a military imperative. It is militarily attractive for each side to attack pre-emptively, and as a consequence it is both attractive and necessary for the other side to do so before it is too late. This situation with a two-sided urge to pre-empt is the most unstable that can be imagined, and to seek to 'balance' forces under these circumstances is to invite disaster. In fact, the more evenly balanced the forces, the more compelling is the urge to pre-empt and the more extreme the instability.

THE IMPERATIVE OF CONCENTRATION

It is easy to see that the division of one's forces into two parts which face the enemy in separate or successive battles implies a direct loss in military capability. If the two parts are equal in size their *combined* military capability is only half of what it would have been if they had fought together. More generally, a division of one's forces into two parts, A and B, results in a loss of military capability given by

$$k.[A+B]^2 - k.A^2 - k.B^2 = 2.k.A.B$$

When combat is subjected to the square law it is therefore essential to concentrate as many forces as possible at the time and place of battle. Nobody would wish to divide his forces into segments unless compelled to do so for other reasons (unrelated to combat proper), for in this case the opponent, by keeping his forces together, would be able to crush the detachments one by one with relatively few losses. If nothing else, substantial superiority in battle after battle ensures a steady improvement in the overall ratio of forces, and, with it, improved prospects of ultimate victory. In this way the opponents mutually compel each other to concentrate their forces.

From this mutual compulsion follows the principle of concentration. This principle applies to the operational level, but its validity hinges on conditions at the tactical level. It is firmly entrenched in orthodox military thinking as the fundamental principle of strategy, simply because conditions similar to the square law are so often applicable. But the principle of concentration is not universally valid. If weapons are very efficient in terms of rate of fire, accuracy and lethality, then the problem is not to *hit* the target but to *find* it – and to find it quickly before the target finds us. Losses, in this case, depend primarily on visibility, and great numerical strength on the battlefield serves no useful purpose. It is rather a handicap because large forces are more difficult to conceal. In terms of overall capability in war it is still important, of course, to have large forces, but in each separate encounter it is the ability to hide which counts. To minimise overall losses it becomes an advantage (other things equal) to disperse forces, instead of concentrating them.[8]

The relationship between firepower and the imperative of concentration is a particularly clear illustration of the way in which models at the different levels fit together: in this case force design (precision fire) affects combat dynamics (invalidating the assumptions of the quadratic model), and combat dynamics, in turn, determines the optimum disposition of forces at the operational level and the general principles of strategy (dispersion instead of concentration). In relation to the alternative models of combat dynamics to be discussed below, it may be noted that the critical point is the fifth of the assumptions underlying the quadratic model. This assumption presupposes that potential targets appear at least as fast as they can be destroyed, so that the critical factor for both combatants is firepower, not detection.[9]

QUEUEING FOR BATTLE: THE FIRST LINEAR MODEL

Lanchester also developed a model to describe ancient warfare in which the weapons of a large army could not be brought to bear because of their limited range or because of terrain configuration, as in the battle of Thermopylae. In this so-called 'first linear model', combat takes the form of a fixed number, n, of separate duels which are allowed to run their course without interference by other combat units. The latter simply wait in line and take over when one of the active units is killed. The best illustration of this mode of combat is the phalanx, as it developed in classical Greece. This was a heavy infantry formation, composed of hoplite soldiers with sword and shield, fighting shoulder to shoulder in a compact rectangular array, *n* warriors across and *m* warriors deep. When two phalanxes clashed, only the soldiers in the first row could engage the enemy, and as the front-row soldiers fell, those directly behind moved up to take their place. As each man needed to be protected on his right flank by his neighbour's shield, it was essential to maintain formation, even though it meant that many soldiers were not actively fighting. The present line-up of forces on the Central Front in Europe provides another illustration of queueing. If there were an attack against a limited sector of the front it would be impossible to engage all the divisions needed for a breakthrough side by side. Due to lack of space, most of the attacker's forces would have to stand by idly in the rear, until front-line units were worn out, and they could take their place.

The queueing model can be regarded as a simple modification of the quadratic model. Here again, the number of Blue casualties per unit of time is determined by the number of Red units engaged, namely, *n*, times the kill rate, k_R, defined as before as the number of Blue units a Red combatant can kill per unit of time.

Queueing for battle
Blue losses: $-dB(t)/dt = n.k_R$
Red losses: $-dR(t)/dt = n.k_B$

The corresponding trajectories in phase-space are straight lines, all with the same slope k_R/k_B as shown in Figure 9.4. Because initial disparities are not magnified in the course of combat there is no particular instability arising from combat conditions at the tactical level. The trajectories are given by:

$$k_R.R(t) - k_B.B(t) = k_R.R(0) - k_B.B(0).$$

As this equation shows, military capability in this case is of the form $k_R.R(0)$. It is the product of quality (the kill rate) and quantity (the number of combat units). Because military capability is determined by the number of combat units itself, not by the square of this number, the model is said to be linear.

The exchange ratio for the queueing model is simply the quotient

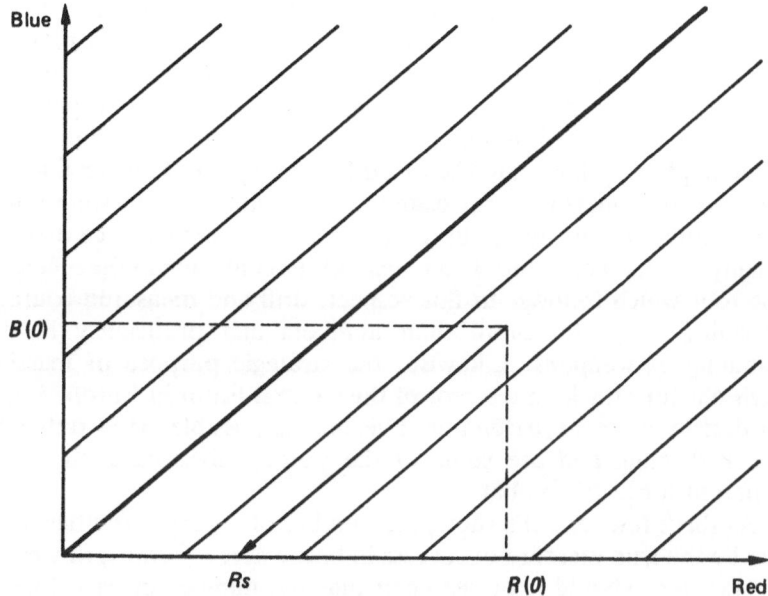

Figure 9.4 Phase-space trajectories for the linear Lanchester models

of the two kill rates. This means that the military value of each combat unit is unrelated to the number of units (friend or foe) present on the battlefield. For this reason the conditions of battle at the tactical level do not impose a need for concentration (or dispersion) at the operational level. Rather, the queueing model describes a dreary process of attrition whose outcome is entirely unaffected by operational manoeuvres and overall strategy. Because skilful operational planning cannot improve the overall exchange ratio, and poor planning cannot worsen it, the disposition of forces at the level of the entire theatre will be determined by considerations other than the need to minimise combat losses. When combat follows the quadratic model the belligerents are forced to concentrate all their forces on crushing the armed forces of the opponent. The theatre of war is then essentially a stage on which the opposing armies seek to outpace and outmanoeuvre each other. When combat follows the linear model, however, strategy and operational art are freed from this constraint, and the deployment of attack and defence forces at the operational level can be determined more directly by the intrinsic political and strategic value of specified parts of the theatre.

In consequence of this, combat in accordance with this linear model may not have decimation of the opposing army as its main objective. 'Linear combat' is not necessarily a process of attrition to be carried to the finish. In many cases, it is only an initial phase of combat, a necessary preliminary to a breakthrough to another, more decisive phase, which could be of a different type (not 'linear'). In the shock of two phalanxes, for example, the point is not to wear down the enemy, man by man, but to deliver a blow of such force that the enemy breaks ranks, and then to exploit his confusion in the mêlée or the rout which follows. In this respect, drill and measured courage probably count for more than numbers and proficiency in the handling of weapons. Likewise, the strategic purpose of massing divisions for attack on a sector of the Central Front in Europe is not to destroy as many defending divisions as possible. It is rather to break through and use some of the waiting divisions for entirely different forms of combat.

As these few remarks show, the number of enemy casualties need not be the true measure of success in battle nor, *a fortiori*, of victory in war. Nor should it be assumed that the number of casualties is primarily determined by combat dynamics and by conditions at the tactical level. In a rout, casualties depend on conditions at the operational level (the ability to block the enemy's escape or to take up pursuit), rather than on combat dynamics in the narrow sense. This all goes to show that a simple Lanchester model is applicable, at most, to one phase in a combat, and not necessarily that phase which is most important in terms of strategic implications. In view of a widespread tendency to derive 'criteria' or 'probabilities' of victory from models of the Lanchester type, it is worth insisting that combat dynamics belongs at the level of tactics, and the question of victory and defeat cannot be discussed meaningfully at that level.

SEARCHING FOR TARGETS: THE SECOND LINEAR MODEL

Lanchester's second linear model was initially developed to describe the effects of indirect (unaimed) fire, as in the case of artillery shelling at random when casualties depend, not only on the intensity of fire, but also on the density of the targets. As Svend Clausen notes, the model may also describe the 'opposite' situation when accuracy and firepower are so high that every target that can be seen can also be destroyed almost instantaneously. In this case, the limiting factor is not firepower but detection, and this depends both on the number of detectors (that is, the number of combat units) and on the number of targets. Modern platforms such as tanks or ships with weapons of high accuracy and lethality, operating in dispersed formation on an extended battlefield in search of enemy units, probably offer the best illustration of this kind of combat.

Let d_R denote the detection rate for the Red forces, meaning the probability that some specific Blue unit is found by a Red unit during one unit of time. The rate at which Red forces find Blue targets (and destroys them) is then given by this detection rate multiplied by the number of Red units, $R(t)$, and by the number of Blue units, $B(t)$, in the area. Likewise for the Blue forces. The model therefore takes the following form:

Searching for targets
Blue losses: $-dB(t)/dt = d_R.B(t).R(t)$
Red losses: $-dR(t)/dt = d_B.R(t).B(t)$

This model is in many ways quite different from the previous one but it gives rise to the same straight line trajectories in phase-space. The exchange ratio is now the quotient of the detection rates, not the kill rates, but like the first linear model it is a constant, independent of the density of forces. The reason is that if one side brings more forces into a given area it increases its chances of finding enemy units, but by providing more targets it also helps the enemy's search in precisely the same proportion. Higher density increases the casualty rate for both, but does not affect the balance of losses. In other respects as well the dynamic properties are exactly as before: military capability is measured in the same way; stability conditions at the tactical level are the same, and so forth. Since the cost of destroying one enemy unit is independent of the disposition of forces in the theatre, this model, like the previous one, describes a pure war of attrition, such as drawn out search-and-destroy operations in guerrilla/counter-guerrilla combat.

ASYMMETRY BETWEEN OFFENCE AND DEFENCE

The models described thus far are not well suited for the analysis of defensiveness at the tactical level and of its implications for stability at this and other levels. The reason is that they cannot do justice to the fundamental distinction between the defensive and offensive modes of fighting. With these models the advantages associated with fighting in the defensive can only be expressed in terms which are ultimately arbitrary and exterior to the model itself, namely, by ascribing a higher kill rate or a higher detection rate to defensive as against offensive combat. This is clearly inadequate.

In the asymmetric models discussed below the distinction between offence and defence is not just represented in the values assigned to the parameters but is built into the mathematical structure of the model. This step has far-reaching implications in terms of the dynamics of combat.

ATTACK ON DUG-IN DEFENCES

In the first asymmetric model to be discussed, Red is assumed to attack with massive firepower along a sector of the front. The Blue forces, not knowing where to expect an attack, are spread evenly and relatively thinly along the entire front, but they have the advantage of fighting from prepared, concealed positions, so that the Red units cannot easily find suitable targets to attack. Red, by contrast must come out in the open to attack, thus providing Blue with many readily identifiable targets. As Svend Clausen suggests, this may be taken as a reasonable description of a massed attack on the European Central Front under conditions as they exist today.[10]

In this model the limiting factor for Blue is firepower, so the Red losses are determined as in the quadratic model. For Red, the limiting factor is the difficulty of finding targets to attack, and the Blue losses are therefore determined as in the second linear model. This results in the following asymmetrical model:

> *Dug-in defences*
> Blue losses: $-dB(t)/dt = d_R.B(t).R(t)$
> Red losses: $-dR(t)/dt = k_B.B(t)$

The trajectories in phase-space are shown in Figure 9.5. They are parabolas, given by the equation:

$$d_R.[R(t)]^2 - 2.k_B.B(t) = d_R.[R(0)]^2 - 2.k_B.B(0)$$

As before, this equation shows that the two expressions on the left

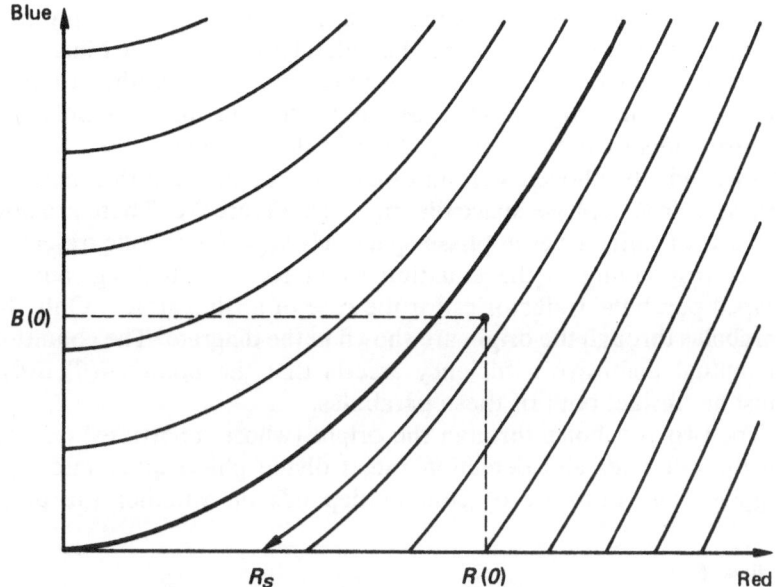

Figure 9.5   Phase-space trajectories for attack on dense, dug-in defences

can be used as measures of military capability. For Blue, the
defender, military capability is a *linear* function of combat strength,
whereas for Red, the attacker, military capability is proportional to
the *square* of the combat strength. In consequence of this, the
behaviour of this model may be described as semi-linear.

STABILITY

In combat against dug-in defences the military capability of given
forces depends on whether they fight in the offensive or in the
defensive. In this case, therefore, it is possible to give a precise
meaning (at the tactical level) to the concepts of offensive and
defensive capability, introduced in connection with the discussion of
mutual defensive sufficiency. If these capabilities are denoted D_R,
D_B, 0_R and 0_B respectively, the condition of mutual defensive
sufficiency.

$$\begin{cases} D_B > 0_R \\ D_R > 0_B \end{cases} \text{ becomes } \begin{cases} 2.k_B.B(0) > d_R.[R(0)]^2 \\ 2.k_R.R(0) > d_B.[B(0)]^2 \end{cases}$$

These inequalities are the conditions of stability which apply in a limited sector of the front when the defender is able to dig himself in, in accordance with the assumptions of the model. According to these inequalities, military stability demands that the initial number of combat units of the defender, ($B(0)$ in the first inequality, $R(0)$ in the second), be above a certain threshold. Again, what this means is best seen from a phase-space diagram as in Figure 9.6. There are now two sets of trajectories in phase-space: U-shaped parabolic trajectories corresponding to the situation when Red is attacking, and C-shaped parabolic trajectories for the case of a Blue attack. Only the parabolas through the origin are shown in the diagram. The condition of mutual defensive sufficiency asserts that the point ($R(0),B(0)$) must lie 'inside' both of these parabolas.

The two parabolas through the origin (whose 'apertures' depend on the kill rates and detection rates) divide phase-space into four regions. The outcome of combat depends on whether the point

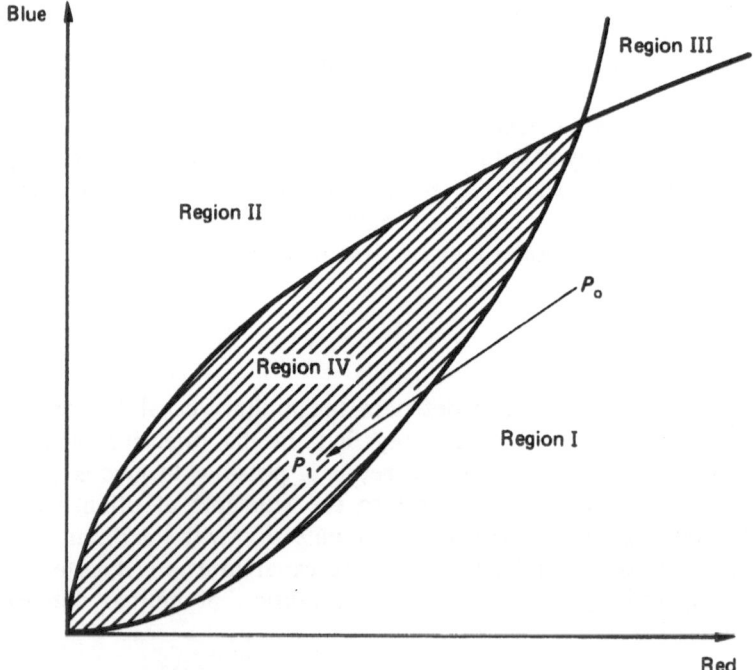

Figure 9.6   Stability and instability regions for the asymmetric, semilinear model

$(R(0),B(0))$ from which combat starts falls in one or other of the four regions:

- Region I corresponds to Red superiority. For points in this region, Red comes out as winner, whoever attacks.
- Region II corresponds to Blue superiority. For points in this region, whoever attacks, Blue wins.
- Region III corresponds to offensive superiority. For points in this region, whoever attacks, wins. Both sides are under extreme pressure to pre-empt.
- Region IV corresponds to defensive superiority. For points in this region, whoever attacks, loses. In this situation of mutual defensive sufficiency it would be folly to launch an attack.

Figure 9.7 illustrates the stability conditions in Region IV. If Red

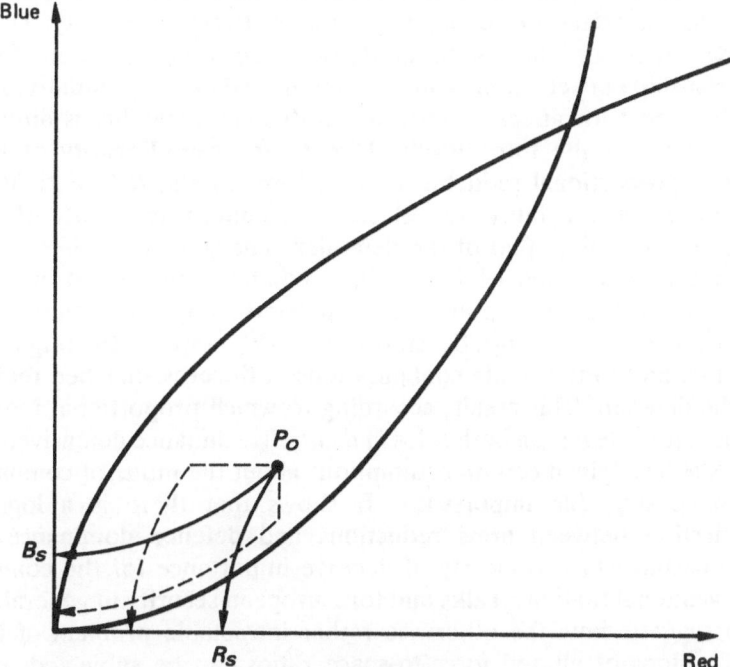

Figure 9.7   Defensive sufficiency for the asymmetric, semi-linear model
$\overline{P_O B_S}$: Red attacks, Blue wins
$\overline{P_O R_S}$: Blue attacks, Red wins
Broken lines: surprise attack

attacks, the trajectory follows the curve P_o-B_s, and Blue wins. If Blue attacks the trajectory is P_o-R_s, and this time Red wins. On a sector of the front where these conditions prevail, both sides have an interest in waiting for the other side to attack. In other words, this is the condition which ensures stability against horizontal escalation. The broken curves in Figure 9.7 illustrate the situation when there is a premium on surprise attack. If this premium is sufficiently large, stability may no longer obtain.

EFFECT OF ARMS REDUCTIONS

All the models considered previously – those at the tactical level, those at the operational level and those at the strategic level – were essentially indifferent to the overall level of forces. In all these models an across-the-board increase or decrease in the forces on both sides has no effect (or no major systematic effect) on stability. These models, therefore, have nothing to say on the crucial question of the relationship between joint arms reductions and military stability. But in the model for attack against dug-in defences, stability is directly dependent on the force levels. Due to the semi-linearity of this model, proportional reductions in the force levels, $R(0)$ and $B(0)$, would enhance stability by reducing the military capability of the attacker more than that of the defender. The same point is evident from a consideration of Figure 9.6. Whatever the initial combat strength of the two opponents, a proportional, across-the-board reduction of forces brings the starting point P_o closer to the origin, to P_1, say, and to the stable configurations in the cross-hatched region of the diagram. This result, according to which proportional reductions in force levels on both sides *in themselves* enhance defensiveness and stability (given certain assumptions about the mode of combat), is of considerable importance. It shows that there is a logical connection between arms reductions and defence dominance, a relationship which is clearly of decisive importance for the coming Conventional Stability Talks and for European security in general. It also suggests how the otherwise rather intractable problem of the implications of altered force-to-space ratios can be subjected to a more rigorous analysis.

It is also an important result because it runs counter to received notions in NATO. It is generally taken to be a fact in the West that NATO's present disposition of forces along the Central Front in

Europe is the minimum compatible with effective defence. Even very large reductions in offence-capable forces on the Eastern side, it is claimed, would not permit NATO to follow suit with more than token reductions, lest the force-to-space ratio become so low as to jeopardise defence. Clearly, this position, if maintained, would preclude negotiated arms reductions. The preceding analysis demonstrates, however, that the presumed need for maintaining a high force-to-space ratio, even as offensive forces on the other side are reduced, rests on a hidden assumption about military doctrine, which amounts, in effect, to ruling out *a priori* the possibility of using forces more effectively by specialising in defensive combat.

SATURATED DETECTION

In combat against dug-in defences, as previously described, the military capability of the attacking forces is proportional to the square of their number, whereas the military capability of the defending forces is proportional to the number of defence units itself. This is the reason why proportional force reductions on both sides (let alone disproportionate reductions on the side which is strongest) facilitate defence and improve stability. For precisely the same reason, of course, the model implies that a proportional *increase* in the forces on both sides favours the attacker: as the force levels on both sides increase, the ratio of the defender's losses to those of the attacker (which is given by the slope of the phase-space trajectories in Figure 9.5) also increases, seemingly without limit. It would seem, therefore, that if the attacker decides to mass forces in a narrow sector of the front, he can wipe out the defence in that sector, virtually without any losses to himself, and this is so even if the defence has been able to mass forces in the same sector at the same rate. While dug-in defences seemed effective and stabilising at the tactical level, they seem to be easy to defeat at the operational level.

In fact, this scenario of a massed attack in one sector to achieve a virtually unlimited exchange ratio hinges on an artefact of the model. The detection of targets was assumed to be the critical factor for the attacker. The rate at which targets are detected is obviously proportional to the number of potential targets, $B(t)$. The model assumes that it is also proportional to the number of detectors, $R(t)$, but this last assumption cannot, of course, hold true in general, however large the number of detectors. There must clearly be a limit, R_L say, beyond which additional detectors are redundant. This would be the case if the defender's combat units are not directly visible, even with detectors, so that the attacker must resign himself to moving foward with no targets in sight, hoping to find some as he moves along. Beyond this point of saturation, more detectors do not uncover more targets, and more attacking units, therefore, do not kill more defenders.

The probability that a Blue unit is detected in a unit of time is $d_R.R(t)$. It increases with the number of Red detectors. Denoting l_R the upper limit reached when saturation sets in, the model can be written as follows:

Saturated detection
Blue losses: $-dB(t)/dt = d_R.B(t).R(t)$ if $R(t) < R_L$
$-dB(t)/dt = l_R.B(t)$ if $R(t) > R_L$
Red losses: $-dR(t)\,dt = k_B.B(t)$
with $l_R = d_R.R_L$

The corresponding trajectories in phase-space are parabolas, prolonged beyond the value R_L by their tangent. Examples of these trajectories are shown in Figure 9.8. In this (slightly) more realistic model, as in the previous one, a proportional cut in all forces, Red and Blue, always leads to a situation of greater stability where defence dominance is enhanced (or attack dominance diminished). This feature is unaffected by the introduction of the limit R_L. On the other hand, the attacker is now no longer able to benefit from massive concentration in a limited sector. Indeed, with these dug-in, well-hidden defences he has largely been deprived of the main asset he can normally count on as attacker: the benefits to be had from covert concentration of forces prior to attack. How this comes about, and what it implies is best studied with a somewhat different model.

ROLL-BACK OF A DEFENCE NET

The last type of combat to be considered is an attack by mobile forces against a defence net, consisting of many small, essentially stationary infantry groups, dispersed and concealed in the terrain. The basic idea of the net is to wear down the attacker in a succession of small engagements which he can easily win, but at a cost which he cannot sustain. Such a net could cover the entire territory, or it may be confined to a belt along the border, whose function is to force the

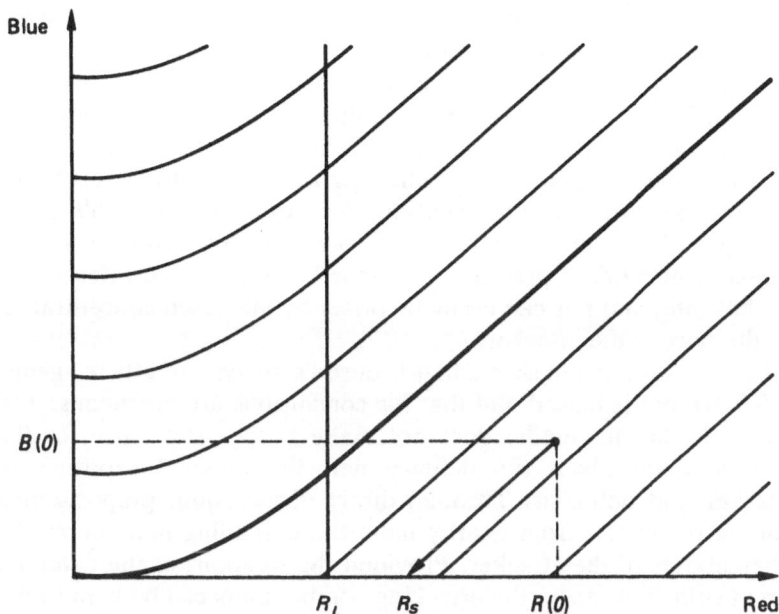

Figure 9.8 Phase-space trajectories for saturated detection

attacker to reveal his axes of attack, to channel and delay the attack while the defence brings in mobile reserves from the rear, and to wear down the attacker's units to a point where the reserves can finish them off.

A defence net, at first sight, is an inefficient way to use the available forces. In actual combat many of the defence units would be bypassed, and would contribute nothing to overall combat strength. Moreover, those units which do give battle would fight at a disadvantage: by keeping his forces together the attacker can meet the defending groups, one at a time, with overwhelming superiority and achieve a very favourable exchange ratio. This line of thinking is at the root of Clausen's analysis of attack against a defence net. Referring to the concept developed by Horst Afheldt, he considers an attack by armoured vehicles against a net of 'technocommandos', equipped with armour-piercing infantry weapons, and dispersed in prepared positions.[11] In his analysis the attacking units move around as a group, engaging the defence squadrons one by one, and it is assumed that in the time from one engagement to the next nothing happens. In each engagement, the limiting factor for the attacker is detection, and the limiting factor for the defender is firepower. Consequently, Clausen's equations for the roll-back of a defence net are exactly similar to those given above for attack on dug-in defences. This implies, in particular, that the military capability of the attacker rises as the square of his combat strength, and the attacker can therefore prevail quite easily, achieving a very favourable exchange ratio simply through concentration. According to Clausen's equations (see, for example, those given above for dug-in defences), the adoption of modern precision weapons by the defence affects only the kill-rate, and this can easily be offset by increased concentration on the part of the attacker.

Closer examination shows that Clausen's analysis of attack against a defence net is flawed and that his conclusions are erroneous. The reason is that his model does not make proper allowance for the decisive point about the defence net: the unequal exposure of attacker and defender. Through direct observation, prepositioned sensors or reports from nearby units the defending units know the whereabouts of the attacker. Provided the weapons of the defender are of sufficient range, the attacking combat units can be kept under constant fire. The attacker, meanwhile, cannot engage any of the defence units until – by chance – he comes close enough to one of them for his sensors to be able to detect it. Then a close combat ensues. This combat may be similar in nature to attack on dug-in defences, or it may not. In any case, the combat situation as a whole cannot be analysed as a succession of close engagements.

In all the previous models it was possible to ignore the space dimension entirely, and analyse combat as if it took place in a single point. For combat against a defence net this is clearly not possible, and geometrical factors such as the distribution of defence forces in the terrain, the movement of attacking units and the ranges of weapons and sensors, must somehow be taken into account. Let it be assumed, therefore, that the defence units are deployed in small groups, scattered evenly in concealed positions over an area, be it the entire territory or a belt, several layers deep, behind the border. Let is also be assumed that each defence unit is fully informed of all enemy movements within reach of its weapons through prepositioned sensors or by other means. Let it be assumed, finally, that defence units can fire at the enemy without revealing their position. This assumption is, of course, critical, and is the main difficulty in designing an effective defence net. The ideal in this respect would be for the defence weapons to be remotely controlled with no two shots or salvoes ever coming from the same position, like a mine or a remotely controlled missile on a one-shot ramp. If the attacker fires back at the launch-point he will hit a discarded canister, not the operator nor the remaining weapon systems.

The problem for the attacking units is that they cannot find and engage the defence units until they come fairly close, and well within reach of the defender's weapons. In this situation, aimed medium- and long-range fire (including precision-guided munitions) are of no use for the attacker. Against widely dispersed and moderately hardened defence units it is also pointless for the attacker to fire at random 'into the brown'. Hence, the only value of the attacker's long-range weapons is that they permit instant concentration of the fire of many units at the point where a defence unit has been found. It is as if the attacking units were equipped solely with short-range weapons, but had essentially unlimited firepower. The situation is similar in nature to an attack by unprotected swordsmen in an area defended by archers in dispersed formation. At close quarters, the archers are helpless and the swordsmen can mow them down with impunity, but during the approach it is the swordsmen who are vulnerable and who cannot strike back. Forced as they are to advance under fire which they cannot return, the attacking units will obviously want to move forward at maximum speed to minimise exposure. They will also want to advance in sufficient strength to ensure early detection and immediate destruction of any defence unit within reach of their sensors. Therefore, as the attacker advances, the ground is swept clear of defence units. All those who have not managed to withdraw at the last minute are eradicated. This means that there is nothing left to do for attack units coming up behind. Altogether, then, the best tactic for the attacker is to place all his combat units in a line, and to press ahead at maximum speed, slowing down only as required to annihilate the defence units encountered on the way.

It is now an easy matter to describe the dynamics of this kind of combat. Referring once more to the attack by Red against Blue, let r_B be the effective range of Blue's weapons (and sensors), and let v_R be the speed at which the attacking Red units push forward. (More precisely, the range, r_B, is to be understood as the range of the defender's weapons, minus the range of the attacker's sensors.) As before, k_B denotes the Blue kill rate, that is, the number of enemy units a Blue unit can kill per unit of time. Finally, let $R(0)$ be the number of attacking units per kilometre along the front, and let $B(0)$ be the total number of defending units behind one kilometre of front in the entire depth, W, of the defence net. $B(0)/W$ is then the density of defence units in the net.

The precise character of the close combat between attacking units and a nearby defence unit after it has been detected is of little consequence. Here we simply assume the worst from the point of view of the defence. First, it is assumed that the defending units are unable to withdraw. This means that the number of Blue casualties per unit of time is simply the number of Blue units initially present in the area swept clean by the Red forces in that time interval. Secondly, it is assumed that the attacking forces are so numerous and powerful that defence units, once found, are wiped out instantly at no cost to the attacker. This means that close combat does not contribute to the Red losses at all. Consequently, the total number of Red losses per unit of time is given by the total number of surviving Blue units within shooting range, multiplied by their kill rate. This results in the following equations:

Roll-back of a defence net
Blue losses: $-dB(t)/dt = v_R.B(0)/W$
Red losses: $-dR(t)/dt = k_B.r_B.B(0)/W$

This model differs markedly from the earlier ones. It is a model of combat dynamics at the tactical level, functionally equivalent to the Lanchester models, but its parameters refer directly to the geometry of weapons interactions and belong 'lower down' in the hierarchy of models, at the level of force design.

Lanchester's own work contains a brief reference to a somewhat comparable situation in which a numerically superior force, initially at a disadvantage, must come to close quarters as rapidly as possible, in order to reap the benefits of numbers and concentration. He describes combat between a large contingent of riflemen and a few machine gunners. In this case, the superior firepower of the gunners weighs in most heavily at large distances when both sides must fire 'into the brown', so that combat follows a linear model. At closer range, where each man and gunner is an individual mark, and many riflemen can target a single gunner, the quadratic model takes over and the advantages inherent in numerical superiority begin to tell. Thus circumstances may arise in which the riflemen either face slow defeat through attrition from a distance, or the possibility of a quick victory if they accept the heavy casualties involved in coming rapidly to close range.[12]

It is at once clear from the above equations that the exchange ratio (the ratio of Red losses to Blue losses) is $k_B.r_B/v_R$. This is a constant which depends solely on the 'technology', not the combat strength of the opponents, and the combat trajectories in phase-space are therefore straight lines, as in Figure 9.4. A suitable measure of military capability for forces deployed in a defence net is $k_B.r_B.B(t)$. For forces attacking such a net, the corresponding measure of military capability is $v_R.R(t)$. It is easy to show that the difference between these two expressions is invariant under combat, as it should be:

$$k_B.r_B.B(t) - v_R.R(t) = k_B.r_B.B(0) - v_R.R(0)$$

Military capability is therefore proportional to the number of combat units, both for the attacker and for the defender, and consequently there is no incentive for concentration. When he confronts dug-in defences, the attacker can minimise his own losses by attacking in great strength, and Clausen implies that this is also the case when attacking a defence net. The above analysis shows that this is not so. The attacker must, of course, gather enough forces to penetrate the net to the desired depth, but, further than this, concentration serves no useful purpose. In short, by organising his forces as a net, the defender can impose a mode of combat which deprives the attacker of his only significant advantage, namely, the ability to increase the military value of each of his combat units by massing his forces at the point of attack.

The military capability of dug-in defences is $k_B.B(0)$. For a defence net it is $k_B.r_B.B(0)$. Both expressions are in the form of a product of the 'quality' of the defence forces by their quantity. In the former case, this quality is simply a question of effective firepower as measured by the kill rate. In the latter case, however, the 'quality' of the defending forces is a product of kill rate and of range, in which these two factors amplify each other. This is the reason why the possibility of setting up net-like defences is so closely connected with the development of modern precision-guided weapons, for these weapons are not simply characterised by a particularly high destructiveness, as Clausen would have it. What is unique about them is that they can achieve high kill rates, even at long ranges.

For the attacker confronting a defence net, the measure of the 'quality' of his forces is simply the speed at which they can advance. Nothing else matters, since it was assumed that the attacker's weapons and detectors are so perfect that the defence units are wiped out instantly, as soon as they come within range of the detectors. This extreme assumption about the capability of the attacker has the merit of laying bare his true weakness: his dependence on speed. This is a decisive weakness. Whereas the 'quality' (the speed) of the attacker's forces is very hard for him to improve, there are many simple and inexpensive ways for the defender to degrade it, for example, by minefields, ditches, and forest belts. Due to the factor v, such obstacles have a dramatic effect on the exchange ratio, on the relative strength of the defensive and on stability.

It is instructive to compare attack against a defence net to 'queueing for combat' as in the first linear model. In fact the model for a defence net can be derived from the queueing model by assuming a kill rate, k_B, for the attacker of v_R/r_B, and a number of duels, n, equal to $r_B.B(0)/W$ (which is the number of defending units whose fire can reach the attacking units).

In sum, the deployment of the defence in a net imposes disadvantageous conditions on the attacker in two respects. First, the attacker is denied the benefits of concentration at the tactical level (as military capability becomes a linear, not a quadratic function of the number of attacking units). Secondly, the conditions of combat are such in this case that the attacker's kill rate is determined, not by the quality of his own weapons, but by factors essentially controlled by the defender: the speed at which the attacking forces can advance through the terrain and the range of the defender's fire.

The present model can easily be extended (though at the expense of simplicity and clarity) to account for the interaction with other arms such as artillery, ground-attack aircraft, and surface-to-air defence. For each additional arm an additional differential equation is needed to describe how that arm declines over time, and an additional term must be added to each of the other equations to describe the effects of its fire. The exact form of such equations will depend in part on assumptions about the way in which combat units distribute their fire over different types of enemy units. It seems that such elaborations of the model will not alter the general conclusions reached above. As long as the basic difference in visibility and vulnerability remains, additional means on both sides should tend, on the whole, to favour the defence. For example medium- or long-range rocket artillery, dispersed and concealed among the other units of the net, could add significantly to defensive firepower, especially if it is combined with a system of remote sensors to provide target data and with minefields and passive obstacles to channel and delay the attack. Such dispersed long-range artillery could also threaten follow-on forces as they advance through gaps where the defence net has been cleared away. In contrast, the attacker's long-range artillery would have to rely on exceedingly complex and expensive types of target-detecting and homing ammunition, or confine itself to un-aimed fire which cannot do much damage.

As the forces composing a defence net cannot be used for attack, the question of bilateral stability at the tactical level does not arise, unless there is a combination of defence nets and mobile, attack-capable forces on both sides. In this case, clearly, a *sufficient* condition for mutual defensive superiority on some limited sector of the front is that the number of mobile combat units in that sector (on either side) does not exceed a certain threshold:

$$k.r.N(0)/v_A$$

where v_A is the speed of the attacker's mobile units, and where k, r and $N(0)$ are the kill rate, the range and the number of units in the defence net on the opposite side. When the number of mobile units is below this threshold, the defence net suffices to hold back an attack. When it exceeds this threshold the attacker has the means to clear a corridor through the defence belt, and the defender must bring in some of his own mobile forces to achieve defence dominance. This brings us 'one step up' to the operational level and to models dealing with stability in terms of theatre mobility and delaying barriers.

For an attacker facing a defence net, speed is a decisive factor: it is imperative at the tactical level in order to minimise casualties and break through the defence belt; it is imperative at the operational level in order to outpace the defender's mobile reserves; and it may also be imperative at the strategic level to ensure a decision before reserves can be mobilised and allies can be brought to bear. Altogether, delay emerges as one of the keys to defensive superiority and military stability at all levels.[13]

Notes

1. See, for example, Anders Boserup, 'Non-Offensive Defence in Europe', in Derek Paul (ed.), *Defending Europe: Options for Security* (London: 1985).
2. See Anders Boserup, 'Mutual Defensive Superiority and the Problem of Mobility along an Extended Front', in this volume.

3.   See Heinrich Siegmann, 'A Simulation of Defensivity', paper submitted to the Pugwash Study Group on Conventional Forces in Europe at its Sixth Workshop held at Altamura, Italy, from 1 to 4 October 1987; and Manuel Goller, 'Brechen oder Halten? Modell einer Linienrverteidigung', Max-Planck-Institut, Starnberg, June 1988.

4.   Frederick William Lanchester, *Aircraft in Warfare: The Dawn of the Fourth Arm* (London: 1916). Reference here is to a reprint of his 'Mathematics in Warfare', in James R. Newman (ed.), *The World of Mathematics*, vol. 4 (New York, 1956) pp. 2138–57.

5.   Lanchester made brief reference to such models. They are worked out more fully in Howard Brackney, 'The Dynamics of Military Combat', *Operations Research*, vol. 7, no. 1 (January–February, 1959).

6.   The most systematic attempt in this direction is Svend Clausen, 'Krig paa formler', *Forsvarets Forskningstjeneste*, June 1987 (Danish Defence Research Establishment, FOFT F–4/1988).

7.   Lanchester, 'Mathematics in Warfare', in Newman (ed.), *The World of Mathematics*, vol. 4, p. 2142.

8.   See Robert Neild, 'The Implications of the Increased Accuracy of Non-nuclear Weapons', in this volume; and Anders Boserup, 'Krigens Vilkaar: Teknologien', *Krig og Fred*, no. 4 (1986).

9.   John Lepingwell, 'The Law of Combat? Lanchester Reexamined', *International Security*, vol. 12 (1987–8) p. 93.

10.   After illustrating the complete impotence of small forces in the presence of overwhelming power, Lanchester noted that this conclusion manifestly does not apply to the case of a small force concealed or 'dug-in', since in this case the larger force is unable to bring its weapons to bear. He did not, however, develop this idea any further. See Lanchester, 'Mathematics in Warfare', in Newman (ed.), *The World of Mathematics*, vol. 4, pp. 2150–1.

11.   See Clausen, 'Krig paa formler', citing Horst Afheldt, *Defensive Verteidigung* (Hamburg, 1983).

12.   See Lanchester, 'Mathematics in Warfare', in Newman (ed.), *The World of Mathematics*, vol. 4, p. 2148.

13.   This chapter was submitted to the Pugwash Study Group on Conventional Forces in Europe at its Seventh Workshop held at Amsterdam from 11 to 13 November 1988.

10 The Defence Efficiency Hypothesis and Conventional Stability in Europe: Implications for Arms Control
Reiner Huber and Hans Hofmann

INTRODUCTION

According to the definition proposed by the Advanced Research Workshop on Modelling and Analysis of Arms Control Problems held in Spitzingsee, in October 1985,[1] the strategic situation between two antagonistic parties is considered to be stable if

● neither side feels compelled to react, on an equivalent footing, to changes in the other side's force posture (for example, inventory increases, modernisation, new options, and doctrinal innovations) in order to maintain its security (Arms Race Stability); and if
● in a given crisis, neither side perceives an advantage from attacking first or, vice versa, both sides consider it more advantageous to react to the other side's attack (Crisis Stability).

In stark contrast to the rest of the world, the European scene since the end of the Second World War has been distinguished by the complete absence of major military conflict, suggesting that a high degree of crisis stability has persisted throughout that region. However, in the same period, the arsenals of the two regional military alliance systems of the North Atlantic Treaty Organisation (NATO) and the Warsaw Pact have expanded considerably. Thus, one is tempted to conclude that arms race instability is the price to be paid for crisis stability, that is, a low probability that a crisis will erupt in war. Yet, a continuation of the ongoing arms competition between NATO and the Warsaw Pact for the sake of maintaining crisis stability seems like an absurd notion, apart from the fact that the economic burden involved will eventually become unbearable. In addition, in view of the ever growing damage potential of the antagonistic arsenals and the associated risk of catastrophic conse-quences in case of a war, a low probability of the occurrence of such a war is considered cold comfort by many.

Thus, one of today's crucial questions is whether and under what circumstances a high degree of crisis stability in Europe can be obtained without a continuous arms build-up. In other words, what are the prerequisites for crisis stability to be conducive rather than detrimental to arms race stability?

MILITARY STABILITY CONCEPTS AND THEIR IMPLICATIONS

Introduction

NATO and the Warsaw Pact seem to belong to diametrically opposed schools of thought on the notion of uncertainty in bringing about military stability, that is, a situation in which neither side may employ its military forces against the other, not even for purposes of coercion. Within the framework of its strategy of flexible response, NATO considers a degree of uncertainty for the success of a Warsaw Pact attack a necessary and a sufficient ingredient for military stability. Flexible response calls for the capability to deny the aggressor success at whatever level of conflict he chooses. Ultimately, he would be confronted with annihilation by the strategic arsenal of the United States.[2] One of its underlying principles is that of conventional sufficiency which implies that, because of nuclear deterrence, military stability can be maintained below conventional parity.[3] In other words, NATO assumes that the Warsaw Pact would attack only if it regarded the degree of uncertainty for accomplishing its operational objectives quickly to be low or, equivalently, the probability for a swift military victory by conventional means to be high. In contrast, the Soviet Union and its allies apparently maintain that military stability requires the capability of 'maximizing the probability that their forces can accomplish their assigned missions in the face of operational uncertainties and enemy countermeasures'.[4] In the context of defensive missions, that is, equivalent to a low degree of uncertainty that the defence will be successful or, equiva-lently, a high probability that the attacker will fail. It will be shown below that this attitude implies conventional superiority. This is also true for the principle of reasonable sufficiency or defence sufficiency recently proclaimed by the Soviets and interpreted by many Western observers as an indication of a fundamental change in Soviet military doctrine.[5] In 1987, the Soviet Minister of Defence, D. Yazov, explained that principle:

> Generally it means having just as many armed forces as is necessary for defence from an outside attack. It means specifically that the personnel of the armed forces, and the amount and quality of means of armed struggle are strictly commensurate with the level of military threat, and the character and intensiveness of the military preparations of imperialism; they are determined by the requirements for assuring the safety of the Warsaw Treaty and for repulsing aggression. At the Political Consultative Meeting in Berlin it was stressed that the armed forces of the allied states are kept in a state of battle readiness sufficient for avoiding a surprise attack. Should they nevertheless be attacked, they will give a crushing rebuff to the aggressor.[6]

It will become evident from our analysis that the capability for a 'crushing rebuff' requires defence superiority. However, given the present Warsaw Pact force structure, defence superiority implies offence superiority even if numerical parity between NATO and the Warsaw Pact were to be obtained. The elimination of offence superiority while maintaining defence superiority necessitates certain structural and equipment changes in the military forces. Thus, in addition to approximate parity, the implementation of the principle of defence sufficiency requires the adoption of defensive force structures incapable of large-scale offensive operations. Otherwise, defence sufficiency is but a euphemism for conventional superiority.

In order to illuminate the implications of these attitudes for crisis and arms race stability, we shall express the issue in a more formalised manner by means of a simple dynamic model of military conflict between two antagonistic parties X and Y. We denote by

$P_{X(Y)}$ = probability that the attacking side $X(Y)$ accomplishes its operational objectives quickly (victory probability);

$W_{Y(X)}$ = probability that the defending side $Y(X)$ is successful (defence probability).

It is self-evident that the probability of a successful attack by one side is equal to the complementary value of the probability for a successful defence by the other, that is

$$P_X = 1 - W_Y \tag{10.1a}$$

$$P_Y = 1 - W_X \tag{10.1b}$$

These equations show that considering a high victory probability $P_{X(Y)}$ as a prerequisite for an attack by one's opponent is equivalent to regarding a low defence probability $W_{X(Y)}$ as sufficient for deterring such an attack. Thus, NATO's concept of military stability implies that, due to the threat of nuclear escalation, a comparatively low probability for a successful conventional defence represents a risk sufficient for an effective deterrence of a presumably risk-averse opponent such as the Warsaw Pact. And because of its high risk aversion, the Warsaw Pact equates military stability with a high probability for a successful defence.

Let us now consider three hypothetical cases of mutual military stability attitudes in the light of possible outcomes of the (initial) conventional phase of a military conflict between the two fictitious parties X and Y:

- Both antagonists X and Y adhere to NATO's stability concept;
- Both antagonists X and Y adhere to the Warsaw Pact's stability concept;
- Side $X(Y)$ adheres to NATO's and side $Y(X)$ to the Warsaw Pact's stability concept.

Stability Under Conditions of Operational Symmetry

Based on the results of suitable dynamic models, R. Avenhaus *et al*[7] and the present writer[8] have described the outcomes of military conflicts in terms of the probabilities of victory P and defence W that result for given states (x,y) of the military capabilities of the antagonists X and Y at the beginning of hostilities. This is illustrated in Figure 10.1 for the first two cases under conditions of operational symmetry for which

$$P_X + P_Y = 1 \tag{10.2a}$$

and because of equation (10.1) it follows that $W_x = P_x$, $W_Y = P_Y$ and

$$W_X + W_Y = 1 \tag{10.2b}$$

Operational symmetry implies identical operational conditions in a qualitative sense, that is, no side has any significant operational or

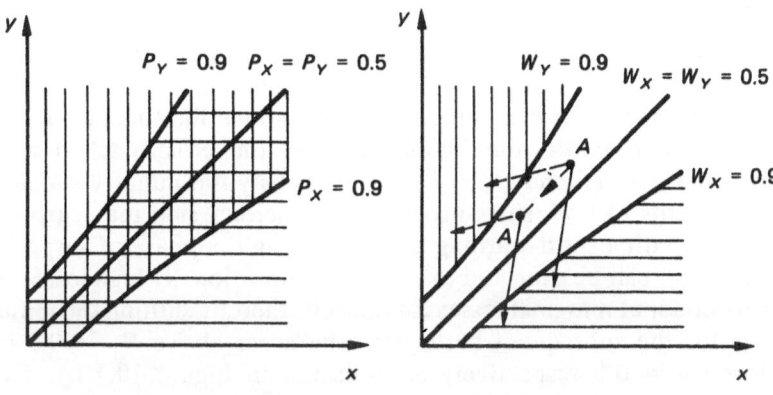

(a) Case 1–*X* and *Y* have NATO-type stability attitude

(b) Case 2–*X* and *Y* have Soviet-type stability attitude

Figure 10.1 Iso-probability functions of victory P_X, P_Y and successful defence W_X, W_Y for opposed sides X and Y

Case 1 is characterised by both sides accepting initial states for which the other side's victory probability P is equal or less than a certain (high) value (e.g. P_X, $P_Y \leq 0.9$) below which an attack is considered too risky. Thus, all states (x, y) within the cross-hatched area of Figure 10.1(a) can be considered as equilibrium points. Case 2 is characterised by both sides insisting on a probability W of successful defence of at least a certain high value (for example, W_X, $W_Y \geq 0.9$). Thus there are no mutually acceptable states (x, y). When initial conditions do not meet the security requirements (for example, points A and A'), either side might in a crisis be tempted to improve its initial position for the anticipated conflict by a preemptive attack as indicated by the arrows in Figure 10.1(b).

tactical advantages over the other regardless of whether it attacks or defends. This is, for example, the case in an open battle or meeting engagement between two forces of similar structure and equipment. In a game–theoretical sense, operational symmetry represents a situation that may be considered as a zero-sum game.[9] Whatever gain one side makes in P or W is a loss for the other. Therefore, operational symmetry precludes high values of W for both sides simultaneously. And, because $W = P$, it implies that a high capability for a successful defence also provides a high capability for being a victorious attacker. This implication is the essence of what Robert Jervis calls the security dilemma: 'Many of the means by which a state tries to increase its security decrease the security of others.'[10]

In the first case, we assume that both sides regard each other as risk-averse to the extent that they would consider an attack only if the probability of P_X and P_Y respectively were sufficiently high, say at least 90 per cent. Thus X could feel safe as long as the capability values of the opponent Y were below the function $P_Y = 0.9$, and Y could feel safe if the capability of X were above the function $P_X = 0.9$. In other words, the initial states in the cross-hatched area between $P_X = 0.9$ and $P_Y = 0.9$ would satisfy the military stability requirements of both sides provided that neither side may surprise the other with a short-warning attack. During the phase of surprise, the attacker can be expected to cause disproportionally high losses to the defender at marginal losses for himself, thereby shifting the initial states for the subsequent battle into the areas above $P_Y = 0.9$ or below $P_X = 0.9$ respectively as indicated in Figure 10.1(b). The stability would be the higher, the closer the state (x,y) approaches balanced situations characterised by the function $P_X = P_Y = 0.5$ of initial states for which the chances of victory are identical for both sides.

The second case assumes that both sides are risk-averse and require at least a 90 per cent probability for a successful defence against an attack by the opponent. The result is depicted in Figure 10.1(b) and shows that there are no mutually acceptable initial states (x,y). Because of equation (10.2b), a high level of security for one side necessarily means low security for the other. Thus the mutual desire for high security becomes the motor for arms competition as either side would continuously strive to shift the initial state (x,y) to the region satisfying its security requirements. But as the reciprocal armaments efforts would prevent just that, either side might in a crisis be tempted to improve its probability for a successful defence in the anticipated conflict through a pre-emptive surprise attack as indicated by the arrows in Figure 10.1(b). Symmetrical arms reductions as symbolised by a shift from point A to A' would not alleviate that problem of crisis instability.

From this analysis we conclude that, without additional assumptions about the capabilities and the structure of the antagonistic military forces, the mutual adoption of a Soviet-style military stability attitude would in fact perpetuate the arms competition and cause crisis instability. In contrast, the adoption of NATO's military stability attitude by both sides would bring about some degree of crisis stability if the capabilities of the antagonistic forces were roughly balanced, if their leaderships were risk-averse, and if short-warning attacks were impossible. However, the crisis stability thus obtained is contingent upon the credibility of nuclear deterrence and would not necessarily lead to arms race stability. This is because a sufficient degree of uncertainty for the attacker to accomplish his objectives requires operational flexibility for the defence. As operational flexibility is proportional to the number of available operational options, both sides would feel compelled to maintain a level of modernisation that assured them against falling behind the opponent's optional capability.

The situation would not be very different in the third case when one of the antagonists pursued NATO's and the other the Warsaw Pact's stability attitude. Suppose X were satisfied with a victory probability P_Y of not more than 90 per cent for Y, and Y required a probability W_Y of at least 90 per cent for a successful defence. In that case, the iso-probability functions $P_Y = W_Y = 0.9$ would coincide, that is, all mutually acceptable initial states (x,y) would be situated on a line. Thus the respective balances would be quite unstable: side X would very likely perceive even a marginal capability increase by Y as an indication of the latter's offensive intentions that call for immediate compensatory actions. And side Y would regard a similar small increase in the capability of side X as a threat to its security that, in a crisis, might cause Y to attack pre-emptively.

The Impact of a Defence Advantage

So far, our analysis has been based on the assumption of operational symmetry, that is, neither side has any significant operational or tactical advantages over the other regardless of whether it attacks or defends. However, given certain conditions, we may assume that the defender does have some advantage over the attacker or, as Clausewitz put it, 'that the defensive form of warfare is intrinsically stronger than the offensive'.[11] For example, a well-prepared defence does have an initial advantage that results, among other things, from concealment, additional protection, and the benefit of the first 'shot' while the attacker has to expose himself in unknown terrain. Subsequently, a mobile defender may attack the aggressor's flanks and rear in a series of surprise blows taking maximum advantage of the familiar terrain.

In formal manner, the defence advantage can be expressed by $W_X > P_X$ and $W_Y > P_Y$, that is for a given capability x or y the probability $W_{X(Y)}$ for a successful defence would always be higher than the probability $P_{X(Y)}$ of being victorious as an attacker. Or, in other words, when confronted with a given military capability of the opponent, one would need a higher initial capability as an attacker for obtaining a given victory probability P than one would need to attain the same value for the probability W of defending successfully against the opponent's attack. Because of equation (10.1), it also follows that

$$W_X + W_Y > 1, \tag{10.3}$$

That is to say, when the defence has an advantage over the offence there is the possibility, at least in principle, for both antagonists to obtain high levels of security. In fact, if the defence advantage were truly significant, conventional superiority might be obtainable for the defender only, that is, conventional superiority would become defence superiority. Also, given some defence advantage, the slopes of the iso-probability functions would change depending on which side attacks and which defends. For example, if Y were the attacker and X the defender, then the slope of the functions P_Y and W_X would become steeper. This is because the defence advantage forces the attacker Y to increase his capability level y in order to maintain a given victory probability over a defender X with a given capability level x. In contrast, the defender X faced with an attacker of a given capability y may maintain a given value W_X at a lower capability level x of his own. The opposite would hold true if X were the attacker and Y the defender. The iso-probability functions W_X and W_Y can be interpreted as representing the partial equilibrium functions for sides X and Y proposed by Kenneth Boulding. Accordingly, the intersection M is identical to what he calls the military equilibrium in a bipolar international system.[12]

These effects of a defence advantage are illustrated in Figure 10.2 for the above-defined cases of mutual stability attitudes. In contrast to the situation of operational symmetry, we now have some areas of mutually acceptable initial states (x,y) not only in the first case but in the other two as well. However, in the second case when both sides adhere to the Warsaw Pact's stability concept, only capability levels above certain minima indicated by point M are acceptable to both sides. For all initial states in the cross-hatched area above M, both sides would have a probability of successful defence of at least 90 per cent. However, unless the defence advantage were truly significant, the cross-hatched area would be quite acute and narrow indicating that the stability of the enclosed capability states (x,y) may still be rather shaky, not least because neither side could ever be certain about the other's true capability level. In that case, stability may only

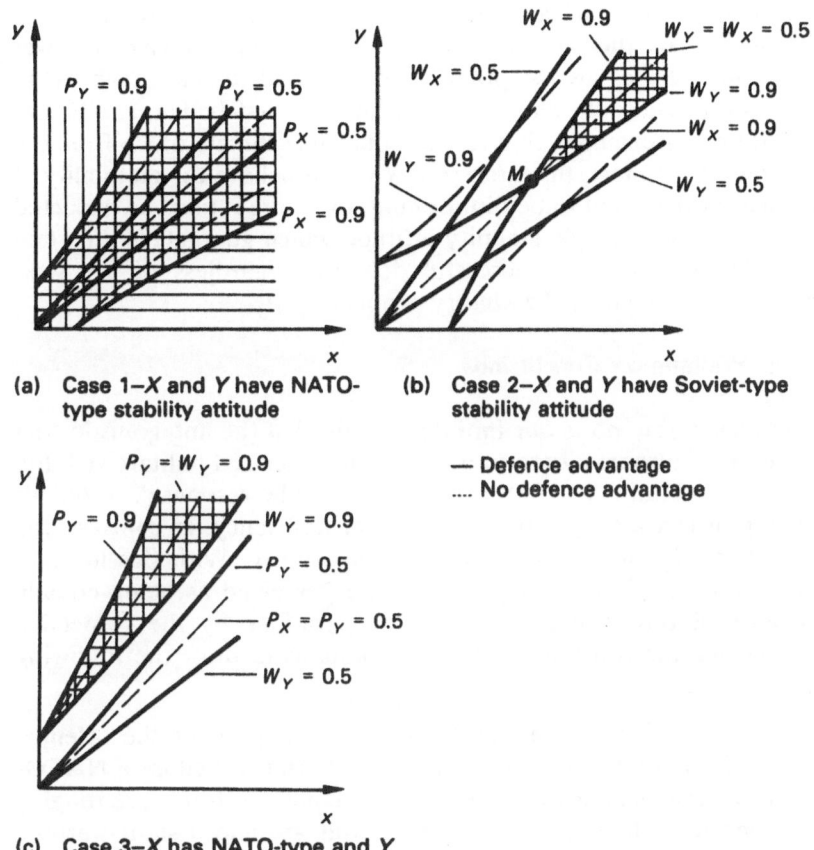

(a) Case 1–*X* and *Y* have NATO-type stability attitude

(b) Case 2–*X* and *Y* have Soviet-type stability attitude

— Defence advantage
.... No defence advantage

(c) Case 3–*X* has NATO-type and *Y* Soviet-type stability attitude

Figure 10.2 The effects of an intrinsic defence advantage

be improved by a mutual reduction of security requirements. For example, if both sides were satisfied with $W_x = W_y \geqslant 0.5$, Figure 10.2(b) shows that the minimum mutual capability state *M* would coincide with the origin $(x=y=0)$. This is but a formal expression of the well-known fact that a high degree of mutual confidence and trust is an indispensable prerequisite for true disarmament.[13]

Similar arguments hold true for the asymmetrical case as depicted in Figure 10.2(c) where side *X* adheres to NATO's and side *Y* to the Warsaw Pact's stability concept, that is, *X* feels safe as long as *Y*'s capability level *y* is below $P_Y = 0.9$, and *Y* as long as *X*'s level *x* is to the left of $W_Y = 0.9$. However, with a view to an erosion of the credibility of its nuclear deterrent, side X may want to revise its assessment of the risk-attitude of side Y and perceive a lower probability P_Y as an acceptable risk-threshold for an attack by Y. This could lead to the elimination of the cross-hatched area in Figure 10.2(c) if, for example, X were to reduce its estimate of P_Y from 0.9 down to 0.5. In fact, the more side X would believe the credibility of its nuclear deterrent to be diminishing, the more it must be expected to adopt a Soviet-style stability attitude which guarantees a certain probability W_X for a successful defence, regardless of Y's risk-attitude expressed by the victory probability P_Y.

Some Preliminary Conclusions

This analysis supports our initial assertion that the antagonistic pact systems in Europe adhere to diametrically opposed military stability concepts. Within the framework of 'flexible response', NATO's concept is characterised by conventional sufficiency. In contrast, the Soviet concept implies conventional superiority. With this in mind and given that force structures may be perceived as being equally capable of offensive and defensive operations by the respective opponents, the foregoing analysis permits us to draw the following conclusions:

- Unless the defence has an intrinsic advantage over the offence, crisis stability may only be obtained if both sides adopt a NATO-type military stability attitude, if the antagonistic forces are roughly balanced, if their leaderships are risk-averse and if short-warning attacks are impossible. However, the arms competition would almost certainly persist, if only because neither side could ever be sure about the opponent's true attitude.

- The more the credibility of nuclear deterrence diminishes for whatever reason, the higher is the likelihood of the adoption of a Soviet-type stability attitude by both antagonists. In that case, some degree of crisis stability may be obtained only if there is a more or less significant defence advantage, and the mutual force capabilities are balanced at sufficiently high levels. But again, incentives for arms competition would persist, if only to be able to maintain the defence advantage.

- Thus, unless preceded by certain structural changes that reduce the offensive capabilities, deep cuts in conventional forces might indeed be destabilising, the more so the less pronounced the intrinsic defence advantage is. This is because, if surprised, a defence thinned by the negotiated reductions would face considerably more unfavourable force ratios at the points where the attacker pursues his main thrusts, especially if the absence of a nuclear threat would permit him to concentrate his forces to a high degree. In addition, since the force reductions would deprive the attacker of some of his follow-on forces, he must be expected to rely even more on a quick initial success by means of surprises.

• As the defence advantage would only benefit a prepared defence, the elimination of surprise attack capabilities is a necessary pre-requisite for crisis stability. However, with a view to an eventual unavailability of nuclear weapons for deterrence, that must be considered as a first step only. This is because time would be less of a critical factor for the attacker who no longer needs to outpace a nuclear reaction by the defender. True stability requires that the area of military capability levels that satisfy the security require-ments of both sides are increased.

PREREQUISITES FOR CONVENTIONAL STABILITY

Introduction

There seems to be but one way for a significant increase in the area of capability states (x,y) that satisfies high security requirements by both sides: redesign of their military forces in a manner that improves the defensive capabilities while simultaneously reducing the offensive potential. There have been several proposals in that direction ever since J. F. C. Fuller published his essay entitled 'Armour and Coun-ter Armour' in 1944.[14] They all suggest that today's 'active' forces, that is, general purpose land forces designed for employment in all combat modes including offence, should be replaced to some degree by 'reactive' forces specifically designed for defensive operations in the terrain prevalent within their respective areas of operation and incapable of incursions into the opponent's territory. Figure 10.3 shows a schematic representation of such a mixed force consisting of reactive and active elements. This nomenclature was originally coined by Saadia Amiel who proposed a land force structure 'of two components: one consisting of defensive combined arms teams, committed to reactive defence where precise and high fire power is at a premium, and the second of offensive combined arms formations

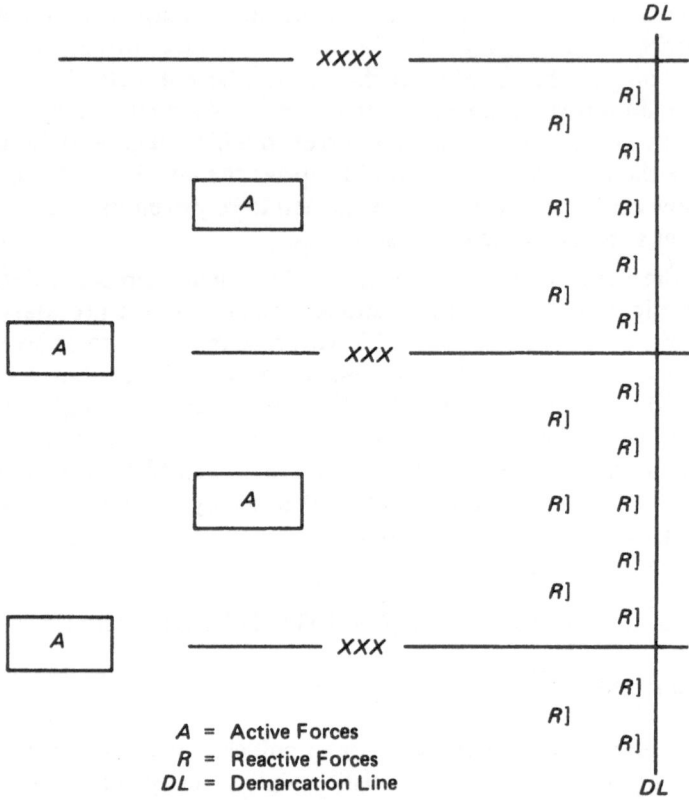

Figure 10.3   Deployment scheme of an active–reactive defence mix

The reactive elements (*R*) absorb the initial onslaught fighting an attrition-oriented delaying battle, thereby providing time for the active elements (*A*) to deploy at the points of the enemy's main thrusts and for counter-attacks into exposed flanks.

where manoeuvrability is at a premium'.[15] The reactive elements would be deployed in some depth next to the demarcation line and, when attacked, serve as a kind of shield absorbing the initial onslaught. They fight an attrition-oriented delaying battle and provide the time required for the active forces to deploy at the points of the enemy's main thrusts and for counter-attacks into exposed flanks.

At a first glance it may seem that the creation of 'reactive' forward defence belts on both sides of the demarcation line – such as the light infantry defence subzones recently proposed by S. J. Flanagan[16] –

would, by itself, improve crisis stability by increasing warning and reaction time for the defending forces. However, the present writers have argued that the replacement of active by reactive elements would make military and economic sense only if the so-called Defence Efficiency Hypothesis (DEH) could be validated.[17] Otherwise, defence capability as well as crisis stability must be expected to suffer as long as there are substantial numbers of active forces left on both sides.

The Implications of the Defence Efficiency Hypothesis

The DEH implies that reactive forces may exploit the intrinsic defence advantage more efficiently, that is, at higher cost-effectiveness than active ones. If this were true, a gradual replacement of active by reactive elements in the opposed forces would have the effects illustrated in Figure 10.4. There, state i may be regarded as representing a situation in which both sides have active forces only. In state k, a certain portion of the active forces has been converted into reactive ones. The functions S_X and S_Y represent the iso-probability functions $W_X = 0.5$ and $W_Y = 0.5$ respectively.

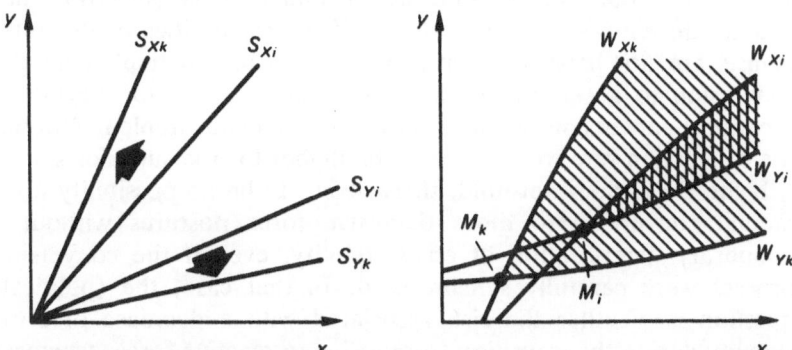

Figure 10.4   Effect of an increasing defence advantage as force structure moves from state i to state k

Subject to the validity of the Defence Efficiency Hypothesis, the transition from state i to k may be effected through a (partial) conversion of general-purpose (active) land forces (capable of defensive and offensive operations) into (reactive) forces specialised in defensive operations (at the expense of offensive capabilities). The Defence Efficiency Hypothesis implies that 'reactive' forces may exploit the intrinsic defence advantage more efficiently than 'active' ones.

As reactive elements would be introduced at the expense of active ones, an increase in defensive capability would automatically be accompanied by a decrease in offensive capability, resulting in an opening of the scissors given by the functions S_X and S_Y and an increase in the 'stable' area enclosed by the iso-probability functions W_X and W_Y. Figure 10.4 also illustrates that a continuation of the mutual conversion process would permit a progressive reduction of military force levels without endangering crisis stability. This is because the minimal force levels M that assure crisis stability become smaller as the degree of force conversion increases, approaching zero when both sides have only reactive forces left. In that case, the functions W_x and W_y would coincide with the coordinate axes y and x, that is, any initial capability state (x,y) would be acceptable to both sides. In other words, a gradual mutual conversion would eventually result in an ultra stable situation in which the armaments level of one side becomes essentially immaterial to the security of the other, thus eliminating the rationale for arms competitions.

A detailed discussion of how a force conversion must be expected to affect the mutual defence capability and crisis stability has been presented by one of the present writers.[18] He concluded that, conditional on the DEH being validated, only a negotiated and synchronised conversion would assure that crisis stability does not deteriorate temporarily. In addition, if the capabilities of the antagonists are not balanced, the superior side has to implement the conversion to a certain degree before the inferior side begins its reciprocal conversion or, in case of a simultaneous implementation, the conversion rate would have to be higher for the superior side.

If the DEH does not hold, there seems to be no possibility for a mutual conversion to more defensive force postures without a temporary deterioration of crisis stability, even if the conversion process were carefully synchronised. In that case, the (political) question is whether the risk associated with a decrease of crisis stability during the transition from active to reactive force structures is worth taking in order to have the chance of eventually arriving at an ultra stable strategic situation.

On the Validation of the Defence Efficiency Hypothesis

While there appears to be sufficient empirical evidence attesting to the intrinsic defence advantage as such, there are only a few instances in military history when circumstances approached those of a well-prepared, reactive-type defence as visualised in most of the afore-mentioned proposals. Fuller's archipelago system of defence, proposed in 1944, exhibits the essential characteristics of a reactive defence.[19] Its underlying rationale is based largely on his observations of some German defensive operations in the North African campaign. Other examples may be the Soviet defences in the Battle of Kursk in 1943,[20] and the German defence during operation 'Goodwood', in Normandy, in 1944.[21] Thus combat simulation experiments represent about the only chance for a systematic investigation of the DEH.

Thus, the development of suitable computer simulation models for testing the DEH on the tactical and operational level has been one of the major efforts within the research programme of the present authors' institute, namely the Institut für Angewandte Systemforschung und Operations Research. The tactical level experiments are to assess the cost effectiveness of reactive elements relative to active ones in the initial battle, and the operational level experiments are to test the interface of reactive and active elements as well as the latters' capability to regain lost territory.

While the operational (corps/army) level model is still under development, several hundred combat simulation experiments involving four variants of active elements and more than ten differently designed reactive modules equipped with presently fielded weapons have been tested in a variety of circumstances by means of the tactical (battalion/regiment) level model BASIS. The latter is a stochastic Monte Carlo-type battle model that permits the closed simulation of battalion-size ground forces defending against a sequence of regimental-size attacker forces accounting for organic as well as higher echelon fire support. It explicitly models each combat vehicle and dismounted infantry down to anti-tank teams of three men. The effects of each shot are simulated including the associated visibility degradation. The experiments take place in digital models of several pieces of real estate in West Germany. The terrain model uses a grid size of 10×10 metres and 10 centimetres altitude resolution and takes account of natural and artificial obstacles and vegetation.[22]

Figure 10.5 shows some selected results derived from the simulation experiments in terms of the relative cost effectiveness of the respective battalion-size reactive modules (on a logarithmic scale) over the relative operational depth required for the attrition of three consecutively attacking Soviet motor rifle regiments in the terrain around Bubach in Bavaria. Cost effectiveness is defined as the ratio

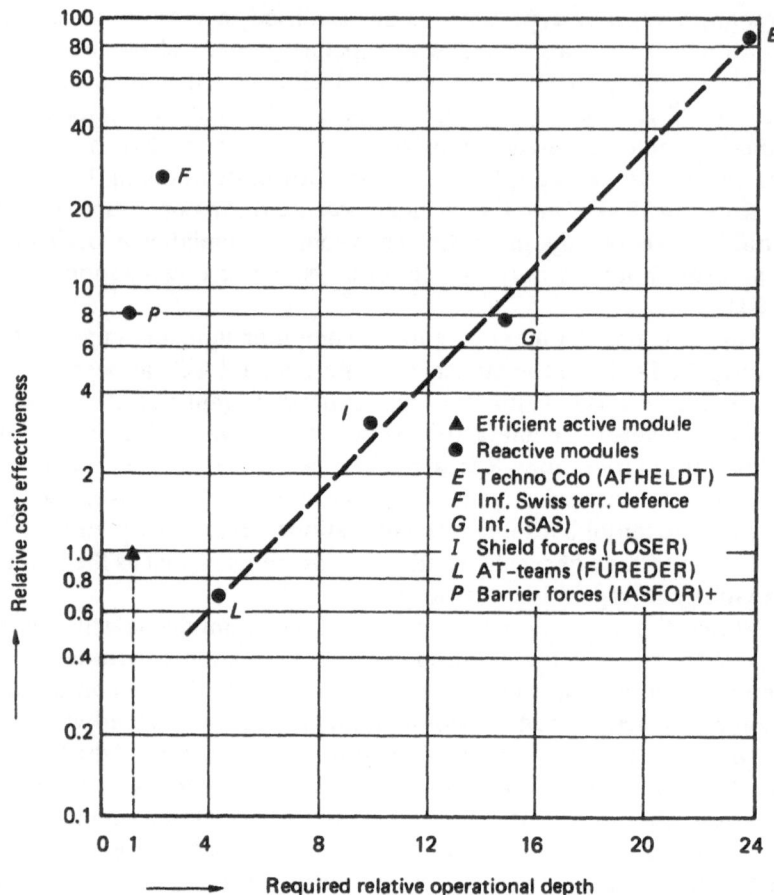

Figure 10.5 Relative cost-effectiveness of battalion-size defence modules over the Relative Operational Depth required for the attrition of three consecutively attacking motor rifle regiments under standard scenario conditions

of effectiveness to investment cost with effectiveness being measured in terms of the relative loss exchange ratio of attacker versus defender. For both, cost effectiveness and cost, the values are multiples of the values that resulted for the active module which turned out to be the most cost effective of the four active variants tested in the respective scenario.

Even though the results pertain to but one of several scenarios, they nevertheless reflect a general trend observed in the tactical level experiments: the validity of the DEH is conditional on the operational depth available to the defender. Suppose the operational depth postulated by the forward defence doctrine for the attrition of three attacking regiments were at most eight times the depth required by the active reference module. In that case, none of the reactive modules *E*, *G*, and *I*, and *L* satifies the DEH. The modules E, G, and I certainly have a higher cost effectivenss than the active reference module. However, their operational depth requirements do not meet the forward defence criterion. Module *L* does so, but merely at about 70 per cent of the cost effectiveness of the efficient active module.

The trendline established by the modules *E*, *G* , *I*, and *L* reflects primarily differences in type and density of anti-armour weapons deployed within the area to be defended. At the upper end, *E* represents infantry teams of little mobility equipped with man-portable weapons and, at the lower end, *L* highly mobile anti-tank teams equipped with self-propelled elevated weapon platforms. *G* and *I* have some mix of both. The direct fire weapon density differs by a factor of more than 4 between *E* and *L*.

In addition to a weapon density 2.5–3 times of that of module *L*, the modules *F* and *P* are distinguished by a high degree of passive protection afforded through strong points (*F*) and field fortifications (*P*) assembled prior to combat from commercially available prefabricated structures stored in the vicinity of their usage. Thus these modules turned out to be fairly robust against preparatory artillery fire as well, the effects of which are not included in the numbers in Figure 10.5.

With a view to the requirement of forward defence, these results support the assertion that, for the presently fielded weapons technology, the validation of the DEH on the tactical level may well depend on the defender's willingness to establish either a permanent system of small and well-concealed field fortifications and/or one that can be put into place on rather short notice together with some barriers designed to channel the attacks. Arguments against fortifications and tank barriers often point out the uselessness of the Maginot Line in preventing the French defeat in 1940. However, as the Maginot Line was largely circumvented by the German invasion coming through Holland and Belgium, that defeat does not disprove the line's utility on the tactical level. On the contrary, the circumvention appears to be quite consistent with what Heinz Guderian had to say about the tactical effectiveness of an uninterrupted line of fortifications against armour. [23] And the operational-level ineffectiveness of the Maginot Line was brought about mainly by the dissipation of 'active' forces, that is, the French tanks which, by the way, were more than equal to those of Germany in numbers and superior in armour protection and gun calibre.[24] It is also worthwhile to note that the organisation of the French defence along the Belgian border in 1940 bears some resemblance to what NATO proclaims to do in the forward defence of Central Europe.

J. Despres suggests that opposition to defensive fortifications in West Germany comes mainly because of their visibility and 'from those who hope for an early demilitarization or elimination of the inter-German border'.[25] However, modern field fortifications and barriers need not be very visible and would have little resemblance to the Maginot Line.[26] But even then, the DEH may yet be falsified by the operational level tests. In that case, modern weapons and communication technology may well hold the key for the DEH being valid when operational depth is lacking. This is why modernisation constraints should be approached with great care in arms control agreements. Unless restricted to systems that are definitely offensive in character, modernisation constraints may indeed become counter-productive with respect to the eventual attainment of a stable military regime in Europe.

Unilateral Validity of the DEH

At least in principle, stability may also be enhanced if the DEH should hold for one side only provided that side is a *status quo* power, that is, its strategic objectives are defensive. In addition, that side must be either *a priori* superior or in a position to field some additional reactive units temporarily in order to compensate for an eventual initial degradation of its defence capability as the unilateral conversion from active to reactive forces commences.

One of the present writers' discussion of the effects of an initial defence degradation is based on the observation that, in order to be operationally effective in a given mission, a certain minimum number of military systems must be available.[27] Thus, depending on the initial force balance, a conversion of active into reactive forces would result in more or less of an initial loss of defence capability until that operational minimum is reached. Given the DEH were valid, the defence capability would increase for degrees of conversion beyond the operational minimum, eventually exceed the original value, and culminate at an upper level beyond which the active forces become too few for restoring the territorial *status quo ante* through counter-attacks. That upper level represents the 'optimal' degree of unilateral conversion that provides the maximum defence capability under the respective circumstances.

Serious objections to a unilateral conversion must be expected, however, both in the case of a balanced situation of the antagonists and also in the case of superiority of the side for which the DEH holds. As one could never be sure about the opponent's true strategic objectives, it seems that the additional fielding, in a balanced situation, of reactive units until their operational minimum is reached would be quite unacceptable to the other side for which the DEH does not hold. This is because a worst-case assessment must cause the latter to consider efficient reactive defences as a means to protect the active forces in their staging areas before an attack. The same argument was voiced by the Soviet participants when they were presented with the idea of 'defensive defence' at the Pugwash Workshop on Conventional Forces held in Pöcking (FRG) in 1985.[28]

In a situation of prior superiority of the side for which the DEH holds, savings in defence expenditures for the sake of aiding a faltering economy appear to be the only incentive for that side to assume a more defensive posture which, if at all, would improve its defence capability only marginally. For this reason, there is hope that the principle of reasonable sufficiency or defence sufficiency does indeed imply a change in Soviet military doctrine aimed at restructuring the military forces in a manner that is less threatening to NATO and satisfies Soviet security requirements at lower costs. With a view to the preliminary simulation results presented in Figure 10.5 and considering the operational depth available to the Soviet Union *vis-à-vis* NATO, this would seem like a logical step to take provided the Soviet Union's economic problems are severe enough and its strategic objectives truly defensive. Thus, we may conclude that a more or less significant validity of the DEH for both antagonists is a necessary condition for eventually bringing about strategic stability between them.

IMPLICATIONS FOR ARMS CONTROL

In a nutshell, the discussion of the security dilemma by Robert Jervis revolves around four worlds which result from the combination of two elementary variables, whether the offence or defence has the advantage, and whether offensive force postures can be distinguished from defensive ones. The results of the foregoing analysis agree with his assessment that what he called the first world (in which the offence has the advantage and force postures are indistinguishable) is 'doubly dangerous' and the fourth world (in which the defence has the advantage and force postures are distinguishable) is 'doubly stable'.[29] However, the analysis presented in this chapter shows that, unless the DEH is valid, there is neither an economic nor a military incentive to adopt distinguishable defensive force postures, and hardly an attractive possibility for a stable transition from the first or the second to the fourth world of Jervis. In fact, if the DEH does not hold, maintaining a high credibility of nuclear deterrence appears to be an indispensable prerequisite for a stable transition to conventional force postures that, by virtue of their structure and deployment, cannot be perceived as offensive.

One of the present writers has proposed five criteria for offensive force postures: a sufficient quantity of weapon and support systems suited for offensive operations; command and control systems suited for the direction of large-scale offensive operations; long-range logistical support; an offensive military doctrine; and appropriate force deployment.[30] There is wide agreement that, today, the Warsaw Pact meets all five of these criteria, NATO the first two partially. Equipment, structure and command and control systems in NATO have evolved as a result of the dilemma of having to defend forward at the demarcation line with comparatively few conventional forces. A simple calculation shows that, even at an overall force ratio of only 1:1, an attacker in Central Europe could attain initial local force ratios of at least 10:1 in his favour at up to ten different points if the defender, not knowing where to expect the attacks, had distributed all his forces uniformly along the demarcation line.[31] In order to cope with such overwhelming local superiorities, the defender must be able to redeploy his forces quickly in such a manner that force ratios sufficient for an effective forward defence are obtained in time.[32] That requires a high degree of mobility for the defence forces and agility of their command and control system, both basic preconditions for offensive operations. However, because of the missing prerequisites, NATO's offensive capabilities are hardly sufficient for any large-scale incursions into Warsaw Pact territory. Yet Soviet thinking seems to assume that these criteria could be met fairly quickly by a determined effort of NATO. Thus, and with a view to the historical experience in the Second World War on the one hand and the Western debate on new operational concepts on the other, it does not appear unlikely that the Warsaw Pact's worst case perceptions of the threat emanating from NATO are somewhat similar to NATO's perceptions of the Warsaw Pact's threat. Also, strictly from the viewpoint of military effectiveness, little credibility may be attached to a concept of linear defence by means of armour which NATO's forward defence concept basically amounts to. In fact, most military experts seem to consider operational depth an essential ingredient for the effective employment of armour.[33] As there is little operational depth available on NATO's territory, seeking that depth beyond the demarcation line might well be considered by Soviet generals as being NATO's better option.

In the light of the preliminary simulation results discussed above, it appears to be quite consistent with the security requirements of the Soviet Union and its allies to attach a high priority to nuclear disarmament. However, for NATO's security conventional arms control is the key issue, including the search for conventional force postures which satisfy the DEH at a comparatively shallow operational depth. The feasibility of such force postures is a necessary condition for nuclear disarmament that meets not only the Warsaw Pact's but also NATO's security requirements. This is why high priority should be accorded to an extensive testing of the DEH by means of computer simulation experiments with emphasis on the defensive potential of barrier systems, new weapons and information technologies. If the DEH can be validated at all, the requisite testing – including the development of accepted simulation models and data bases – will certainly take some time. This aspect is emphasised by Andrey Kokoshin in the context of a recent discussion on conventional arms reduction.[34] He points out that the differences in the force balance assessments between East and West might be caused, to a large degree, by differences in the respective assessment models. Thus he argues that one important task is either to develop a joint assessment methodology or to become acquainted with the differences so that both sides can agree to apply the requisite corrections. Until then, the following actions are considered to present the minimum requirements for avoiding a deterioration of the stability of the military situation in Europe:

- Maintenance of a sufficient capability for theatre nuclear systems beyond artillery range.
- Reduction of surprise or short-warning attack capabilities (for

example, constraints on the deployment of armour, military exercises, the storage of supplies and offensive combat support equipment in forward areas, and improved surveillance capabilities).

● Establishment of a rough balance by a build-down to agreed upon ceilings for those force components that represent the backbone of the offensive options.

● Selected restrictions on the modernisation of offensive force components in order to pave the way for the transition to more defensively oriented force structures. These restrictions must not impede the meeting of the first requirement and the development of defensive force postures that satisfy the DEH.

● Development of a joint assessment methodology including the dynamic models required for systematic tests of the DEH on a broad scale.

Notes

1. R. Avenhaus, R. K. Huber and J. D. Kettelle (eds), 'Systems Analysis and Mathematical Modelling in Arms Control', *Operations Research Spektrum*, Band 8, Heft 3 (1986), pp. 129–41.
2. West German Defence Ministry, *Weissbuch 1985: Zur Lage und Entwicklung der Bundeswehr* (Bonn, 1985), p. 29.
3. West German Defence Ministry, *Weissbuch 1970: Zur Sicherheit der Bundesrepublik Deutschland und zur Lage der Bundeswehr* (Bonn, 1970), p. 29.
4. A. Liebermann, 'Pre-Arms Control Assessment of the Strategic Balance: The Impact of Objectives and Approaches', in R. Avenhaus, R. K. Huber and J. D. Kettelle (eds), *Modelling and Analysis in Arms Control* (Berlin and Heidelberg, 1986), pp. 151–69.
5. See B. E. Trainor, 'A Strategic Shift Observed in Moscow', *International Herald Tribune*, 8 March 1988.
6. *NOD: Non-Offensive Defense*, February 1988.
7. R. Avenhaus, J. Fichtner and R. K. Huber, 'Conventional Force Equilibria and Crisis Stability: Some Arms Control Implications of Analytical Combat Models', paper presented to Workshop on Supplemental Ways for Improving International Stability, held in Laxenberg, September 1984.
8. R. K. Huber, 'Uber strukturelle Voraussetzungen für Krisenstabilität in Europa ohne Kernwaffen: Eine systemanalytische Betrachtung', *Operations Research Spektrum*, Band 9, Heft 1 (1987), pp. 1–12; and R. K. Huber, 'On Structural Prerequisites to Strategic Stability in

Europe without Nuclear Weapons: Conclusions from the Analysis of a Model of Conventional Conflict', *Systems Research*, vol. 5, no. 3 (1988).

9. See S. J. Brams, *Rational Politics: Decisions, Games and Strategy* (Washington, DC, 1985).

10. See Robert Jervis, 'Cooperation Under the Security Dilemma', *World Politics*, no. 30 (1973), pp. 167–214.

11. C. von Clausewitz, *On War* (edited and translated by Michael Howard and Peter Paret) (Princeton, New Jersey, 1984), book 6, chapter 1.

12. See K. E. Boulding, *Collected Papers, Volume 5: International Systems: Peace, Conflict Resolution, and Politics* (Boulder, Colorado, 1975), p. 365.

13. See also Jervis, 'Cooperation under the Security Dilemma', p. 176.

14. J. F. C. Fuller, 'Armour and Counter Armour, Part 3: Defence Against Armoured Attack', *Infantry Journal*, May 1944, pp. 39–44. See also H. W. Hofmann, R. K. Huber and K. Steiger, 'On Reactive Defense Options: A Comparative Systems Analysis of Alternatives for the Initial Defense Against the First Strategic Echelon of the Warsaw Pact in Central Europe', in R. K. Huber (ed.), *Modelling and Analysis of Conventional Defense in Europe* (New York, 1986), pp. 97–140.

15. S. Amiel, 'Deterrence by Conventional Forces', *Survival*, March–April 1978, pp. 58–62.

16. S. J. Flanagan, *Arms Control and Stability in Europe: An American Perspective* (Wildbad Kreuth, 1987).

17. R. K. Huber and H. W. Hofmann, 'Gradual Defensivity: An Approach to a Stable Conventional Force Equilibrium in Europe?', in J. P. Brans (ed.), *Operational Research '84* (Amsterdam, 1984), pp. 197–211.

18. Huber, 'On Structural Prerequisites'.

19. Fuller, 'Armour and Counter Armour'.

20. See Andrey Kokoshin and Valentin Larionov, 'The Confrontation of Conventional Forces in the Context of Ensuring Strategic Stability', chapter 5.

21. See J. F. C. Fuller, *A Military History of the Western World* (New York, 1956), p. 569; and Sir Hugh Beach, 'On Improving NATO Strategy', in Andrew Pierre (ed.), *The Conventional Defense of Europe* (New York, 1986), p. 177.

22. For a description of BASIS see Hofmann *et al*., 'On Reactive Defense Options'.

23. See Heinz Guderian, *Die Panzerwaffe* (Stuttgart, 1943), pp. 215–18.

24. See Fuller, *A Military History of the Western World*, p. 390.

25. J. Despres, 'Politico-Military Assessment', in Huber (ed.), *Modeling and Analysis of Conventional Defense in Europe*, p. 194.

26. See also Beach, 'On Improving NATO Strategy', pp. 177–8.

27. Huber, 'On Structural Prerequisites'.

28. See also Guderian, *Die Panzerwaffe*, p. 216.

29. See Jervis, 'Cooperation Under the Security Dilemma', p. 211.

30. Huber, 'On Structural Prerequisites'.

31. See Huber, 'Über strukturelle Voraussetzungen für Krisenstabilität'.
32. See also West German Defence Ministry, *Weissbuch 1970*, p. 29, item 25.
33. See, for example, Guderian, *Die Panzerwaffe*; and E. Manstein, *Verlorene Siege* (Frankfurt, 1959).
34. A. Kokoshin, 'Militärpolitische Aspekte der Sicherheit in den Ost-West-Beziehungen', *Politik und Zeitgeschichte*, November 1987, pp. 45–53.

Part III
Defences on Land

11 An East–West Negotiating Proposal

Albrecht von Müller and Andrzej Karkoszka

The *détente* of the 1970s had a short life-span largely because it did not prevent the increase of military threats in Europe. A peaceful transformation of the East–West conflict requires the elimination of these threats. Now both alliances seem to be heading in the right direction. But their evolving positions are characterised by a high degree of traditional thinking and lack of innovation. There is imminent danger of again getting trapped in the quarrels typical of the Mutual and Balanced Force Reduction talks (or 'mutual force reduction' talks, as they are called in the East).

The time is ripe for a joint East–West proposal for a conventional arms control regime in Europe – a modified, more fundamental approach to increasing stability. And here we modestly offer the first such proposal.

Conventional stability exists only when the robust defence capabilities of each side clearly exceed the offence capabilities of the other. To create such a regime of 'mutual defensive superiority', the rewards to attack and pre-emption must be systematically reduced, and the structural advantages of the defender must be exploited in order to compensate for the residual advantages of the attacker.

Understanding the basic differences between attack and defence in conventional warfare is crucial if they are to be decoupled. The attacker has the advantages of pre-emption, surprise, and concentration. In order to exploit these advantages he must be able to strike deep in enemy territory, pre-emptively, and he must have strategic (long-distance) mobility, since such forces require heavy mechanised units with extensive logistic support. The defender has the advantage of fighting on well-known and possibly even prepared territory. In order to compensate for the factors of surprise and concentrated attack he must have mobility as well, but not on a strategic scale.

Because asymmetries exist in the NATO and Warsaw Pact forces, cuts by equal numbers would make the situation even worse for the inferior side. Nor would cuts by equal percentages bring about balance. Only asymmetric cuts leading to equal ceilings would provide symmetry and balance. But even this is not enough. Stability can only be achieved if the reductions are combined with an emphasis on the dominance of defence.

The emerging NATO position, which is to propose equal ceilings just below the present NATO force levels, does not fulfil this criterion. It does not change the force characteristics; it provides no hedges against the single most destabilising factor, deep-strike capabilities; and it is probably not even negotiable because it requires the Warsaw Pact to cut by half while offering only cosmetic changes in present NATO forces.

The most advanced Warsaw Pact offer to equalise the various weapons categories by asymmetric cuts is also insufficient. It, too, fails to hedge against deep strike. Leaving aside the fact that it would be likely to trigger another decade of data quarrels, it would, at best, achieve a balance at lower levels. But lower numbers *per se* do not increase stability. They may even decrease it by facilitating preemption and attack missions.

Our proposal would selectively cut down weapon platforms (tanks, helicopters, aircraft, and so forth) that have attack and penetration capabilities, deep-strike capabilities, and other offence-oriented components. Equal ceilings for these are placed at roughly 50 per cent of the present force level of the inferior side. At the same time, both sides would be allowed to maintain or even acquire as much passive munition and barrier technology, anti-tank and anti-aircraft weapons, and 'close interdiction' systems (with ranges shorter than 50 kilometres) as they deem necessary against the remaining threat.

A stability-oriented arms control regime for Europe, between the Atlantic and the Urals, should comprise for either side:

- a low ceiling for main battle tanks: for example, 10 000, combined with a density limit of, say, 500 tanks per 10 000 square kilometres;
- a comparably low ceiling for heavy artillery (with calibres above 100 millimetres) and rocket launchers, again combined with a density limit that denies offence-capable concentrations;
- a low ceiling for strike aircraft and armoured helicopters, perhaps 500 each;
- a range limitation, for example, 50 kilometres, for conventional rockets and all other unmanned weapon systems;

- geographic limits on ammunition stockpiles, perhaps no closer than 150 kilometres to the border, and a ban on forward deployed mobile bridging equipment;
- logistic infrastructures that require frequent back-up by immobile service stations and other installations.

These are drastic changes that require sensible paths for the transition. But they are necessary for conventional stability, that is, to render strategic offences unattractive, impotent, and self-defeating. They should be attractive politically because they imply deep cuts on both sides – approximately 50 per cent for NATO and 80 per cent for the Warsaw Pact – which looks much better than 5 per cent for the West and 50 per cent for the East, and would also save enormous resources in the long run.

Retaining nuclear weapons as a factor of ultimate incalculability is probably unavoidable for the time being. But in a regime of conventional stability, the nuclear component could be substantially reduced. A limit of 500 warheads for land- and air-based systems in Europe might be sensible, with only 100 of them on rockets, and these of less than 500 kilometres range in accord with the Intermediate Range Nuclear Forces Treaty. The arms control proposals for Europe should be complemented by such a nuclear component.

It would be futile to negotiate extensively on marginal improvements of conventional stability while force modernisations which have drastically destabilising effects continue. At the same time, neither side would want its force modernisations to be held hostage to progress at the negotiating table. These two demands add up to a new double task for modernisations: they must make sense without arms control, but they should also be compatible with it.

Fortunately such modernisation options exist. Again, the idea is to exploit the differences between attack and defence. An example is the idea of combining a network of dug-in seismic sensors with mobile mines. The sensors would help the mines find their targets but would not be destroyed in the blasts – a great cost saving. At the same time, the very structure of this system prevents it from being used for offensive purposes.

It would be preferable to create conventional stability exclusively through selectively disarming components capable of offence. But to the extent that modernisations are unavoidable, they must be politically controlled and tuned to the goal of increased stability.

Since 1986 the Warsaw Pact as such and several of its leaders individually have repeatedly advocated purely defensive military doctrines. In the Brussels Declaration of December 1986, NATO also subscribed to the goal of eliminating capabilities for invasion and far-reaching offence. This verbal consensus is something, but much more initiative and creativity is required in order to transform it into an arms control agreement.

In 1987, high-ranking Soviet officials invited four Western experts – Frank von Hippel of the United States, Anders Boserup of Denmark, Robert Neild of Great Britain, and Albrecht von Müller of West Germany – to write a position paper on how a conventional arms control regime could look. In response, General Secretary Mikhail Gorbachev not only welcomed the overall approach outlined above but also stated:

> We see the way to secure reasonable sufficiency in this: that the states would not possess military forces and armaments above the level that is indispensable for an effective defense, and also in this: that their military forces have a structure that would provide all necessary means for repulsing potential aggression but at the same time would not permit them to be used for the unfolding of offensive missions.

This statement is important for three reasons: first, it shows that the Soviet 'new thinking' is also valid for conventional defence and does not brush aside even such fundamental changes as we have proposed. Second, it clearly defines the criterion of 'reasonable sufficiency', which was somewhat vague before. Finally, Gorbachev no longer advocates changes only in doctrinal issues and force levels but, explicitly and for the first time, in the very structure of conventional forces.

This is a promising signal indicating that the Kremlin leadership welcomes a serious, comprehensive approach to stabilising the conventional realm. Now the conventional arms control process must be accelerated and made concrete. A new approach has been developed which would drastically increase conventional stability. In addition, it would save valuable resources for both sides. It is now the task of the politicans to put this option into effect against what is likely to be substantial institutional inertia and resistance.[1]

Note

1. This chapter was previously published in *The Bulletin of the Atomic Scientists*, September 1988. Permission to reproduce it here is gratefully acknowledged.

12 A Possible Stable Configuration of Warsaw Pact–NATO General Purpose Forces After Radical Reductions From the Atlantic to the Urals

Alexander Konovalov

Let us take a look at how a military structure of opposing forces in a most critical region of Central Europe could be constructed (see Figures 12.1 and 12.2). The line of direct confrontation between the Warsaw Pact and NATO in this region extends for 780 kilometres. According to the proposal of former West German State Secretary of the Ministry of Defence Andreas von Bülow, it can be divided into thirteen defensive sectors each with a frontage of 60 kilometres.[1] The entire border strip is designated as a special zone with a depth of 150 kilometres. This creates thirteen defensive sectors of 60 × 150 kilometres in frontage and depth respectively. It is held to be advisable inside each sector to create three subsectors each 50 kilometres deep.

Now we can suggest the composition of exclusively defensive and counter-offensive forces for each of the thirteen defensive sectors. The numerical ceilings and space concentrations will primarily concern the counter-offensive potential. Each of the thirteen basic defensive sectors can contain one motorised rifle or tank division or an equivalent force. The figures show a scheme with nine motorised rifle and four tank divisions.

The first subsector (0–50 kilometres from the border) of each defensive sector can contain up to one tank or motorised rifle battalion from its division force; the second (50–100 kilometres) can have up to one tank regiment or motorised rifle regiment and the third (100–150 kilometres) can contain the remaining divisional units.

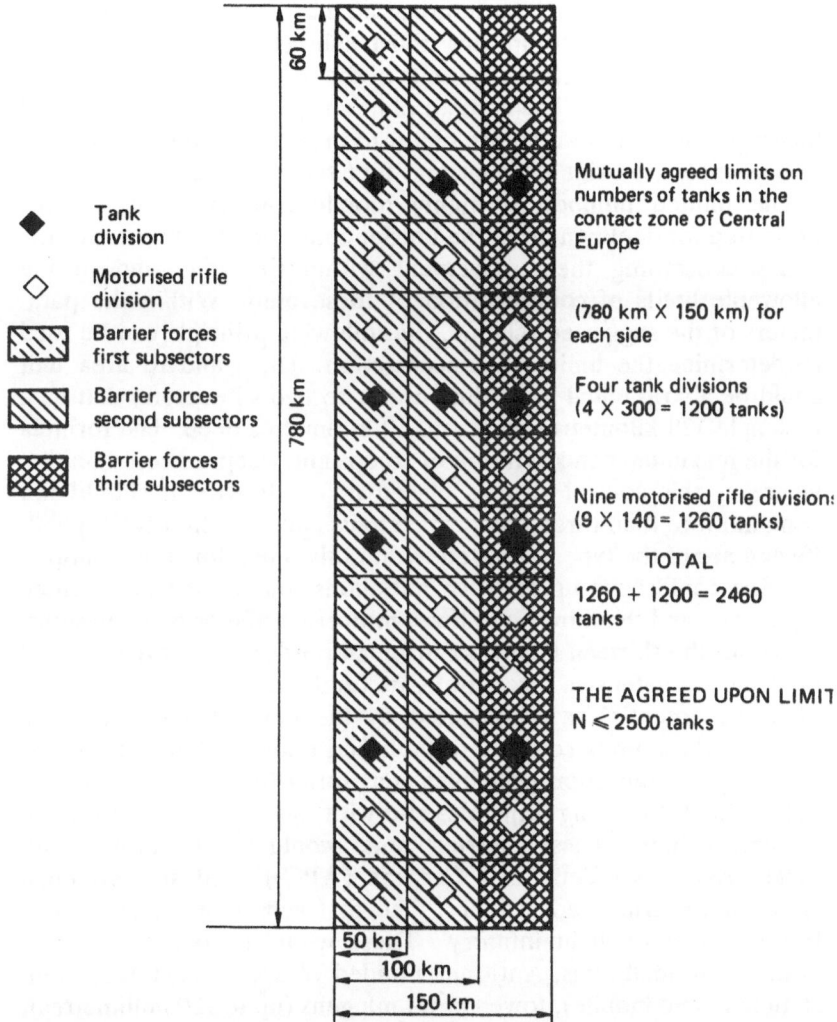

Figure 12.1 Composition of counter-offensive and barrier forces in the contact zone for NATO and the Warsaw Pact in Central Europe (for one side)

In the second subsector the deployment of a certain agreed upon number of tactical non-nuclear surface-to-surface missiles with a range of up to 50 kilometres and Multiple-launcher Rocket Systems (MLRSs) can be permitted. In the third subsector – a strictly limited number of large-calibre (over 100 millimetres) armoured self-propelled artillery and armoured attack helicopters would be allowed. Also allowed in this third subsector would be mobile bridging equipment within the limits of the regular divisional (tank or motorised rifle) level. The number of infantry fighting vehicles should be limited in the 150-kilometre sector to the number of these vehicles in nine motorised rifle divisons and four tank divisions.

The limited number of highly mobile counter-offensive units permitted for deployment in the border zone completely changes the issues concerning their movement within the zone and of the allowable limits of concentration of these units. Within the parameters of the suggested scheme, the following principle can be used to determine the limits of concentration. The standard area unit could be a strip equal in its dimensions to two subsectors, that is, a rectangle 120 kilometres wide and 50 kilometres deep. The formula for the maximum concentration of troops and weapons could be that wherever this control strip is placed in a subsector it should not contain more than three units which have counter-offensive capability and are of the type permitted within this subsector. For example, the first could have up to three battalions, the second up to three regiments, and the third up to three motorised rifle or tank divisions.

Within the thirteen Central European border sectors there could be deployed defensive forces not structurally related to the counter-offensive potential as represented by the motorised rifle or tank division; these units could be established under different organisational and mobilisation principles (territorial forces), with substantially different training principles and with significantly less stringent numerical limits. These 'barrier' units would not be armed with tanks, Armoured Personnel Carriers (APCs), and self-propelled large-calibre armoured artillery. The first subsector would deploy barrier forces of 'light infantry'. These would be permitted to be armed with small arms, Anti-tank Guided Weapons (ATGWs) (both stationary and mobile), towed anti-tank guns (up to 120 millimetres), stationary air defence systems and Command, Control, Communications and Intelligence (C^3I) systems.[2]

The role of the last category of military equipment under these new conditions would be greatly increased. Receiving constant, adequate information about the military activity of the opposite side in the region of direct contact, and proper safe communications with one's own units is one of the necessities of ensuring stability. Therefore, in the border areas it is not only counter-productive to limit these

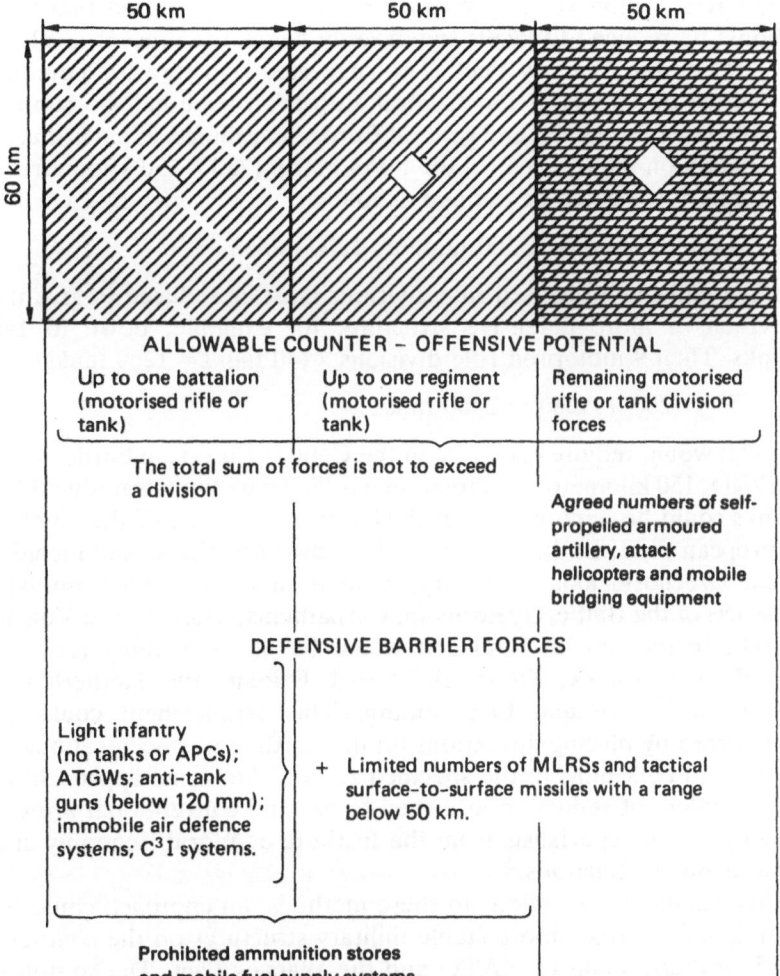

Figure 12.2 Possible structure of one defensive sector in Central Europe

systems, but it would also serve a useful purpose to forbid, near the border, the deployment and use of equipment for electronic counter-measures intended to disrupt the communications and monitoring systems of the opposing side.

So far we have not touched on the issue of how many tanks exist in any particular East European country or in the military districts of the Soviet Union at the present time. But it is proposed that they should be reduced to levels significantly below the present levels of NATO or the Warsaw Pact. The suggested structure for counter-offensive potentials in the Central European zone can be supplemented by subregional and general European ceilings, in particular, a limitation on tank forces. In the border strip (780 × 150 kilometres), the tank force could be defined as follows:

4 tank divisions × 300 tanks = 1200 tanks.

The tanks in motorised rifle divisions must be reduced to give the divisions a more defensive structure, for example, down to 140 tanks. Then 9 motorised rifle divisions ×140 tanks = 1260 tanks.

Σ = 1200 + 1260 = 2460 tanks

This would require each side in the Central European border zone of 780 × 150 kilometres not to exceed a 2500 tank limit. Another 3000 tanks could be deployed by each side in a rear area of the Central European zone. In the case of the Warsaw Pact, this would include: East Czechoslovakia, Hungary, Poland, and the Soviet military districts of the Baltic, Byelorussia, Carpathians, Moscow and Volga–Urals. In the case of NATO it would be: the remaining territory of West Germany, Denmark, Great Britian, the Netherlands, Belgium, France and Luxembourg. This arrangement could be reinforced by placing limitations on the maximum concentrations of tanks and tank units in the specified zones. The 'zonal approach' to the problem of reduction in armed forces and conventional arms is used by Soviet specialists from the Institute of World Economy and International Relations.[3]

With some modifications to these methods, an approach could be developed to create more stable military structures on the southern and northern flanks of NATO and the Warsaw Pact. The southern region poses the most difficult geostrategic problems. In the Balkans Warsaw Pact countries have common borders with NATO countries between Bulgaria, Greece and Turkey; and in the Transcaucasus there is a land border between the Soviet Union and Turkey. Considering also the fact that the Soviet Union shares an even longer border with Iran, with whom it has had an agreement from 1921, and the difficult terrain, a zonal approach similar to the one in Central Europe would prove more troublesome and may not be useful. Still it could be feasible to isolate a 360 kilometre-long corridor within the Transcaucasus military district which borders on Turkey. Here something similar to the first subsector of Central Europe could be established. This would involve a depth of 50 kilometres and a limit of no more than six tank and motorised rifle battalions deployed simultaneously, and a space concentration limit as given earlier with a total limit on tanks not to exceed 150 in the subsector. In total, the mobile counter-offensive forces of the Soviet Union and Turkey in this region could be limited to a total potential of five motorised rifle divisions and one tank division, or:

$$5 \times 140 + 1 \times 300 = 700 + 300 = 1000 \text{ tanks}$$

Taking into account the extended border with Iran, the Transcaucasus military district could be permitted an additional counter-offensive potential equal to one tank or two motorised rifle divisions, or roughly another 300 tanks which would be deployed at a great distance from the Turkish border. Turkey in turn could balance this with additional equipment in its southern Asian territory. From this we get a total of 1300 tanks within the Transcaucasus military district.

On the southern flank of the Warsaw Pact, Bulgaria shares borders with Greece and Turkey. The land border stretches for approximately 500 kilometres. For geostrategic reasons the zonal approach in this area (primarily on the Bulgarian–Greek border) would be difficult to implement. It would be worthwhile to isolate the northern territory of Greece and the western part of Turkey with the permissible limit on tanks of 1000 and the geographical breakdown as follows: 650 tanks in northern Greece and 350 in western Turkey. This tank potential would be balanced by the tank potential of Bulgaria (1000 tanks). Greece's remaining tank potential would be balanced by Romania (550 vehicles), while the Turkish potential (excluding that which was allotted previously for Soviet and Bulgarian borders) would allow for another 1000 tanks to be balanced by the tank potential of the Odessa military district (another 1000 tanks). Thus, the tank forces on the southern flank for both sides (Bulgaria, Romania, the Soviet military districts of the Transcaucasus and Odessa, and Greece and Turkey) would be as follows:

$$\Sigma \text{ Warsaw Pact} = N_{TC} + N_o + N_B + N_R$$

where: N_{TC} = number of tanks in the Transcaucasus military district
N_O = number of tanks in the Odessa military district
N_B = number of tanks in Bulgaria
N_R = number of tanks in Romania

Σ Warsaw Pact $= 1300 + 1000 + 550 + 1000 = 3850$ tanks

Σ NATO $= N_G + N_T$

N_G = number of tanks in Greece, $\underset{\text{(N. Greece)}}{650} + \underset{\text{(S. Greece)}}{550} = 1200$

N_T = number of tanks in Turkey, $\underset{\text{(Soviet border)}}{1000} + \underset{\text{(Bulgarian border)}}{350} + \underset{\text{(Central Asia)}}{300} + 1000 = 2650$

Σ NATO $= 1200 + 2650 = 3850$ tanks

Under this scheme both sides are left with 'southern rear areas'. On the NATO side, these are Spain, Portugal and Italy; for the Warsaw Pact, Kiev and North Caucasus military districts. At this point it would be appropriate to introduce a certain asymmetry into the NATO–Warsaw Pact balance, taking into consideration the significant imbalance on the northern flank, where as stated by the West German expert von Bülow, 115 Norwegian tanks confront 2300 tanks deployed on the territory of the Leningrad military district.[4] Of course, it should be recognised that the Soviet Union has taken some unilateral steps to reduce its military potential in the north-western part of the country. As Eduard Shevardnadze stated on 20 March 1989, at a Moscow meeting with the Deputy Prime Minister and Minister of Foreign Affairs of Finland, P. Paasilo, in recent years the Soviet Union has removed from the north-western part of the country a total of 700 tanks and at present there are no large tank formations employed in the area.[5]

The presence in a future confrontational structure of regional imbalances on the flanks does not, in our opinion, signify a rejection of equal, overall, European ceilings for NATO and the Warsaw Pact. Concerning this issue we would like to suggest the following distribution of tanks. The Kiev and North Caucasus military districts would have 1500 tanks and Spain, Portugal and Italy would have 2050 tanks. In turn, the Leningrad military district would have 650 tanks and Norway would have 100. This gives a total ceiling on tanks for Europe as follows:

Σ Warsaw Pact $= 2500 + 3000 + 3850 + 1500 + 650 = 11\,500$

Σ Nato $= 2500 + 3000 + 3850 + 2050 + 100 = 11\,500$

A similar approach could be used for other elements of the counter-offensive land forces, infantry-fighting vehicles, APCs and large-calibre self-propelled artillery.[6]

Notes

1. Andreas von Bülow, 'Conventional Stability, NATO–WTO: An Overall Concept', Hearings before the US Congress, House of Representatives Armed Services Committee, Washington, DC, 7 October 1988.
2. This question can to an extent be solved through the machine gun/artillery divisions which are being formed in the Soviet Union; these make up a third type of divisions in addition to the two main types, motorised rifle and tank, whose structure significantly changes with the elimination of a tank regiment from the Table of Organisation.
3. O. Amirov, N. Kishilov, V. Makarevsky and Yu. Usachev, 'Problems of Reducing Military Confrontation', *Disarmament and Security: IMEMO Yearbook 1987* (Moscow, 1988), pp. 450–54.
4. Von Bülow, 'Conventional Stability', p. 39.
5. *Pravda*, 29 March 1989.
6. The proposals in this chapter were first introduced in a paper by the author entitled 'Problems of Ensuring Stability During a Reduction of Armed Forces and Conventional Arms in the WTO and NATO' (1989). It was widely discussed by Soviet military and civilian specialists, and by foreign colleagues at various forums.

Part IV
Air Forces

13 Airpower and Conventional Stability
Carlo Jean

THE AMBIGUOUS ROLE OF AIRPOWER IN STABILITY

The role of airpower in the context of conventional stability is difficult to evaluate in an unequivocal way. On the one hand, it acts as a stabilising factor while, on the other hand, it can destabilise. The inherent characteristics of flexibility and mobility of airpower can favour the attack or the defence. Furthermore, it is noteworthy that air forces are means of firepower and not of territorial occupation. They are a source of destruction not of invasion, unlike ground forces such as tanks, armoured vehicles, artillery and other weapons associated with the provision of continuous fire-support required in defence and above all in the attack. A number of simulations has shown that with regard to breakthrough of a defensive position artillery has an effective ratio of 5:1 compared with airpower. Of course, air forces are an extremely important component of any operation whether defensive or offensive.

DEFENSIVE ROLE

The defence is bound to await the attack's initiative. The 'warning time' which allows the defence to react to a surprise attack is more and more reduced and uncertain. The defence is effective only if it has enough time to occupy defensive positions, or to organise itself and dispose its forces in a sufficient concentration against those of the enemy. Therefore it is logical to equate the actions of air forces with those of the 'covering forces'. Air forces have high, structural capability of concentration, allowing a rapid opposition to surprise attack. This compensates for the slow movement of ground forces of the defence. Thus air forces constitute a real counter-surprise factor, favouring the defence rather than the attack. This is also due to the fact that the effectiveness of air forces is greatest against exposed, mobile and concentrated targets, rather than against dispersed defensive units, well concealed and protected. Counter air units are, on the other hand, more effective when deployed in static defensive positions, rather than moving as is necessary during an attack.

In the initial phase of a war, only air forces have the capability, by timely intervention along the line of contact or in depth against the attacking forces, to provide the time needed for slower ground forces to reach defensive positions from their peacetime locations. In the case of defensive sectors which are geographically detached, for which an interior-line manoeuvre would be too difficult or impossible, air forces intervene to provide defence coherence. A sufficient availability of airpower would for this reason ensure conventional stability at a lower level of ground forces, reducing at least partially the minimal level of forces necessary to guarantee a reasonable defensive capability. Otherwise, paradoxically, such a conventional stability could be achieved only with levels of ground forces superior, even in relevant terms, to those of the attacking forces. As a matter of fact, the defence should have in any sector the capability of counter-balancing the potential of the attack, no matter what kind of model of stability is considered. The existence of powerful air forces thus overcomes the geographical limitations on manoeuvre and provides compensation for geographical imbalance – thereby enhancing stability. Any model of simulation, and above all the so-called dynamic models which take into account the possibility for the attack to manoeuvre to achieve the required mass for a breakthrough, can easily demonstrate this fact.

THE ROLE OF AIRPOWER IN OFFENSIVE OPERATIONS

Clearly air forces are essential for any successful operation, including an offensive one, to delay the deployment of the defending units to their advanced positions, to wear down the defences in depth and to perform follow-on forces attacks. A remarkable advantage, moreover, though maybe not as decisive as that attained in the Middle East War of 1967, can be achieved through a surprise air attack aimed at the destruction of the bulk of enemy air forces in their bases. This advantage is structurally destabilising because it inevitably brings with it the temptation to carry out a pre-emptive surprise attack against air bases, which are a weak point in any air force structure. Nevertheless, this potential advantage could be considerably reduced by various means. These include air defence, active and passive protection of air bases, the introduction of Vertical or Short Take-off and Landing (VSTOL) technology, or of aircraft needing shorter landing and take-off space, and the allocation of resources for rapid runway repair. A relevant factor influencing the importance of airpower in the offensive is the existence of mobile air-defence units deployed with the attacking columns. The density and effectiveness of this system is a clear indicator of an offensive structure. Without such projection the attacking columns could not move forward without being subjected to considerable attrition by the air forces of the defence. Mobile anti-air systems should therefore be considered in conjunction with reductions in air forces.

Compared with artillery, air forces are more helpful to defence than to attack. In fact only artillery has the capability of providing the firepower and the persistence necessary to break through the enemy's lines. A breakthrough as such requires great firepower capability against the advanced elements of defence, on the front line more than in depth. In a Close Air Support (CAS) role tactical aircraft are increasingly replaced by combat helicopters and/or multiple launch rocket systems. Conversely, in the interdiction role airpower could never presumably be overtaken by ballistic or cruise missiles (with conventional warheads), given the greater flexibility and payload of the manned aircraft.

DIFFICULTIES IN EVALUATING THE IMPACT OF AIRPOWER

The effectiveness and role of air forces for the ground battle is extremely difficult to evaluate, not so far as CAS is concerned, which can be considered a flying artillery, but with regard to the interdiction role, whether at close range or in depth. It is therefore difficult to develop trade-off models between ground-attack aircraft and artillery or armoured forces. To this uncertainty may be added the fact that at the onset of operations, if there is no immediate threat to the defensive positions, the bulk of all air forces is supposed to be employed in the air battle against the opposing air forces, rather than in supporting the land battle. Nevertheless, should the need occur, the air forces can be diverted from the air battle and be concentrated against the enemy's columns moving towards their assembly areas for the attack. To sum up, the correlations between air and land battle add a remarkable degree of uncertainty to the assessment of the impact of airpower on the latter. The impossibility of correctly evaluating this aspect could hinder or block the negotiations about the more dangerous ground systems and forces.

There are two other factors which cause great difficulties in evaluating the role of air forces in conventional stability negotiations in Europe. The first is the great mobility possessed by air forces which raises the possibility that they could be transferred from 'out-of-area' to the area dealt with in negotiations. The second derives from the fact that almost any combat aircraft has a multirole capability, even if its combat characteristics are optimised for one mission. Thus with minor and relatively simple adjustment of onboard avionics combat capabilities are easily modifiable. It is therefore misleading to speak of offensive or defensive aircraft when both have mixed capabilities.

Any negotiating stance concerning airpower's impact on conventional stability in Europe will thus turn out to be really complex and highly sophisticated. First, it will be necessary to account for the air forces of the two alliances both outside and inside the Atlantic to the Urals (ATTU) area. Given that the primary objective of negotiations on stability is that of making surprise attacks impossible, it is evident that parametrical indicators should be adopted, relating to the deployment of air forces. In practice, more consideration should be given to air forces deployed at bases from which it would be possible to launch a direct attack on the other side than to those air forces deployed at bases requiring a further preventive redeployment forward. This would apply to air forces located in the ATTU area, but perhaps also to those deployed outside this area. For instance, different indicators may be applied to aircraft deployed 250 kilometres and/or 500 kilometres from the boundaries of the two blocs; to those deployed inside the ATTU area but beyond that range; and to those deployed outside the ATTU area.

In the second place, a negotiation will have to take into account all kinds of combat aircraft, including interceptors, reconnaissance (RECCE) squadrons easily convertible to ground attack role, and Electronic Support Measures (ESMs). It is therefore necessary to elaborate relevant parameters to assess the ground-attack capability of various kind of aircraft. Factors to be considered would take into account:

● Weapon platform: weight, speed, manoeuvrability, power, and so on. The most relevant factor is the net weight or Basic Mass Empty (BME), from which aircraft performance can be extrapolated, even if such an indicator may vary according to technological innovation.

To avoid the possibility of a race towards weight reduction with the aim of achieving unilateral advantages, it might be necessary to connect this parameter to other factors and prepare a periodic re-examination of the categorisations.

● Mission characteristics: guidance and attack systems, mission-load, range of action in various configurations and so on.

This constitutes an extremely complex task, which might be assigned to a group of experts from both blocs.

A third problem is posed by naval air forces. Even if they are an integral part of naval forces, which have been excluded from negotiations, it is inevitable that in the long term they will be considered. In any case, it is noteworthy that conventional stability in Europe, defined as no surprise attack and no large-scale attack, is only marginally influenced by any existing naval air power capabilities. This latter is now less relevant than in the past because of the vulnerability of the aircraft carrier, because an ever increasing proportion of embarked aircraft are committed to interception and because of the lesser capability (in terms of payload, survivability and range) of such aircraft compared with land-based ones. One possible measure relating to naval airpower could consist of an offset between units of naval airpower and a given number of interceptors deployed in zones, where a dangerous power projection by such naval air units can be considered a real possibility.

A fourth approach which could be extremely useful for the drawing up of criteria by which to arrive at definitions of stability structures in the field of air forces is the classification of air bases and agreements on their numbers, their deployment in terms of their distances from the frontiers and their characteristics. Focus on infrastructure would permit the definition of objective parameters for stability values much better than if aircraft only were considered. Each base would have to be considered according to its operational potential (the number and type of aircraft which could be deployed and the number of sorties which could be generated), or according to its logistic capabilities, even if a reliable evaluation of these latter would understandably present special difficulties and uncertainties. For example, classification of bases into categories should consider:

● number, length, width, and Load Classification Number of runway;
● reception capacity;

- number and quality of shelters for aircraft and personnel;
- ammunition and fuel stores;
- type and number of hangars;
- road, rail and pipeline links.

Classification of air bases could furnish useful elements for developing stability models for air forces, linking them strictly to the objectives of negotiations on conventional stability in Europe. These problems are undoubtedly difficult and complex. But we have to cope with them to give the negotiators concrete parameters of evaluation and to avoid the situation where the currently opposed positions on airpower could block the negotiating process. It is clear that it is preferable to solve the theoretical problems concerning airpower before entering the actual negotiations.

CONCLUSION

Airpower, through its flexibility and its greater effectiveness in a defensive land battle than in an offensive role (with the exception of the more dynamic phases of exploitation of a breakthrough), is a substantially stabilising element; or at least it is less destabilising than armoured forces and their supporting artillery. Certainly airpower can also play a very important offensive role, which however could be neutralised by the withdrawal or reduction of armoured forces and their associated artillery.

The role of airpower in a defensive land battle is first of all in the area of counter-surprise. Airpower thus is countering the principal advantage of the attacker: the possession of the initiative, which enables him to concentrate his forces at the required time and place. Furthermore, air forces can offset at least in part geographical separations which might hinder movements on interior lines. The purpose of negotiations on conventional stability is not only to reduce armaments, but also to increase security. Therefore, it is not possible to move towards major reductions in airpower if first the much more destabilising asymmetries in ground forces have not been eliminated.

Air forces constitute also a destabilising factor because of the intrinsic advantage of surprise attacks against enemy air bases. Such instability could be reduced however by active and passive measures and by the employment of aircraft which can be less dependent on sophisticated bases (such as VSTOL aircraft or aircraft capable of taking off and landing from motorways). The destabilisation potential of air forces, in other words, could be decreased by reducing their vulnerability.

The mobility of air forces requires that they should be considered on a global basis rather than only within the ATTU area. It is necessary to apply, for the conventional stability evaluation, differential ratios according to the distances of the deployments of air units from the frontier between the two alliances. It is clear that, in terms of stability, an aircraft located in Central Europe has a different impact from one deployed in Siberia or in the United States.

All aircraft have a multirole capability, even if they are different. It is necessary to arrive at a classification by categories – as much as possible based on objective and verifiable data and conversion factors between one category and another – to determine their weight in a potential land battle. It is necessary also to take into account the interrelationship between the potential air battle and the potential land battle.

Interesting possibilities arise for an objective evaluation of airpower from an analysis of the capability of air bases. In the next decade air bases will still constitute a determining factor in airpower. Being earthbound they are a more stable and objective factor through which to give a stability evaluation. It is a subject to analyse at a technical level, in order to identify reasonable solutions acceptable to all the negotiating parties.

Without the prior solution of these problems, the inherent difficulty of evaluating the impact of airpower on conventional stability carries the risk of blocking any possibility of agreement. The Conventional Stability Talks (CST), even though limited to the most important land weapon systems, still have too many elements of the greatest complexity and difficulty. The Western policy of postponing to a second round of CST the question of airpower is justified by the fear of causing a blockage in negotiations, and not from any wish to gain unilateral advantages, which would not make any sense. It is motivated also by the fact that in the airpower field the imbalances between the two alliances are not as great as those relating to ground forces and that, even if airpower is a means of destruction, it is certainly not a means of invasion. In any case, after the beginning of the first phase of CST, the above-mentioned technical problems concerning airpower could be reviewed by task-forces of experts from the two alliances.[1]

Note

1. This chapter represents the personal views of the author. It was submitted to the Pugwash Study Group on Conventional Forces in Europe at its Seventh Workshop, held at Amsterdam from 11 to 13 November 1988.

14 A Zonal Approach to the Neutralisation of Airpower in Europe

Anders Boserup and
Jens Joern Graabaek

INTRODUCTION

The first official recognition of the need to base future European security on defensive principles is to be found in the Warsaw Pact's Budapest Address of June 1986. In this statement the member states rightly identified airpower as an important aspect of conventional arms reductions in Europe. In addition to reductions in ground forces they proposed (as a start) a 25 per cent reduction in the 'tactical strike aviation' of the two alliances.

Mutual reductions in 'tactical strike aircraft' is basically a sound idea. Shifting the balance in favour of defensive systems which would not be reduced (such as fighter aircraft designed and equipped for air defence only and surface-to-air missiles) would facilitate active defence of one's airspace and impede offensive counter-air operations, thereby enhancing overall strategic stability. But this approach has some practical shortcomings as well. Agreement on the definition of the categories of aircraft to be reduced and on the numbers to be reduced and on the numbers to be permitted on each side would not be easy in view of the great differences in the capabilities of air forces of both alliances and in the tasks assigned to them. Differences in ground-based air defence systems may also have to be taken into account in determining mutually acceptable ceilings for specific types of aircraft. Nor is it clear that the proposed cutback in tactical strike aircraft would significantly reduce the incentives and the capabilities for pre-emption. In any case aircraft are easily redeployed, and numerical ceilings, to be effective, might have to be global. This would introduce endless new complications, not related to the European scene. Accordingly, as a supplement (or as an alternative) to reductions in the numbers of aircraft, consideration is given in this chapter to the possibility of a zonal arrangement which would impose

159

restrictions on airfields (and other associated measures) as a means for thinning out tactical aircraft near the border and for impeding offensive air operations.

DESTABILISING EFFECTS OF AIR POWER

Strategic and tactical air forces are a destabilising and highly escalatory element of current military postures in Europe. Aircraft and their ground-based infrastructure are vital military assets, critical for the support of operations on land and in the surrounding seas. The achievement of air superiority, at least in certain areas, is vital for an aggression. Suppression of enemy air power is therefore a top priority for both sides from the very first phases of a conflict.

As air assets are both potentially threatening and relatively vulnerable they rank high on the list of targets for, and as means of, pre-emption. In a war in Europe, whether it results from an act of calculated aggression or from a pre-emptive attack undertaken in despair, it can safely be assumed that the opening moves would include massive counter-air operations to cripple the enemy's air force before it could strike back. The military benefits from successful offensive counter-air operations are such as to create strong incentives for pre-emption. As aircraft stationed outside the initial combat zone can be relocated at short notice, the pressures for pre-emption are not confined to the immediate combat zone. Compelling military logic would encourage both sides to extend offensive counter-air operations to a much wider area. One of the most important tasks in the search for more stabilising and less escalation-prone military structures in Europe is thus to eliminate these incentives for pre-emption and for horizontal escalation.

To increase stability and promote mutual defensive superiority in relation to air warfare, the main objectives should be

- to reduce or eliminate the incentives and the capabilities for pre-emption, thereby enhancing crisis stability and reducing the pressures for horizontal escalation;
- to impede deep penetration of enemy airspace and the achievement of air superiority beyond the front line;
- to strengthen the defensive at the expense of the offensive capabilities of the air forces.

AN 'OPEN AIRFIELDS ZONE'

The mainstay of a zonal approach would be measures to prevent the military use of airfields within the zone. There is no need to prohibit airfields altogether (nor would this be possible in practice). Instead, airfields within the zone would be 'open' or 'demilitarised': hardened shelters and military equipment and installations, such as repair facilities, would be prohibited, fuel stores would have to be unprotected above ground and of a size corresponding to normal civilian requirements, and no fuel pipelines or bunkers for other military purposes, including ammunition depots, would be allowed within a certain distance from the airfields. Prepared areas for landing strips, heliports and other facilities usable for air force purposes would also be prohibited within the zone. To prevent circumvention there should be limitations on air refuelling equipment and training.

As shown in Figure 14.1, such a zone could consist of several bands, extending a few hundred kilometres on each side of the border – with due regard for the existence of national air forces. In the first band (extending, say, 100 kilometres behind the border) all airfields would be 'open' in the foregoing sense. There would be no hardened shelters or protection facilities at all, and in peacetime only fighter aircraft and surveillance and early-warning aircraft would be allowed to operate in this zone. In the next band, 50 kilometres wide or so, there would be some agreed number of shelters, but only for fighter aircraft strictly for defensive counter-air (interceptors), and some hardened fuel stores, but of limited size. Pipeline fuel systems would be prohibited in this zone. In the third band, extending, say, from 150 to 250 kilometres from the border there would be some agreed ceiling on the number of hardened shelters and on the number of tactical aircraft (fighter-bombers) which could be stationed in this area in peacetime.

An 'open airfields zone' is mainly a way of extending early-warning time in order to support strong air defence systems on both sides of the border. More precisely, the main purposes of such an arrangement are

● to increase early-warning time on both sides;
● to facilitate air surveillance beyond the border;
● to impede the build-up of offensive air force assets near the border;

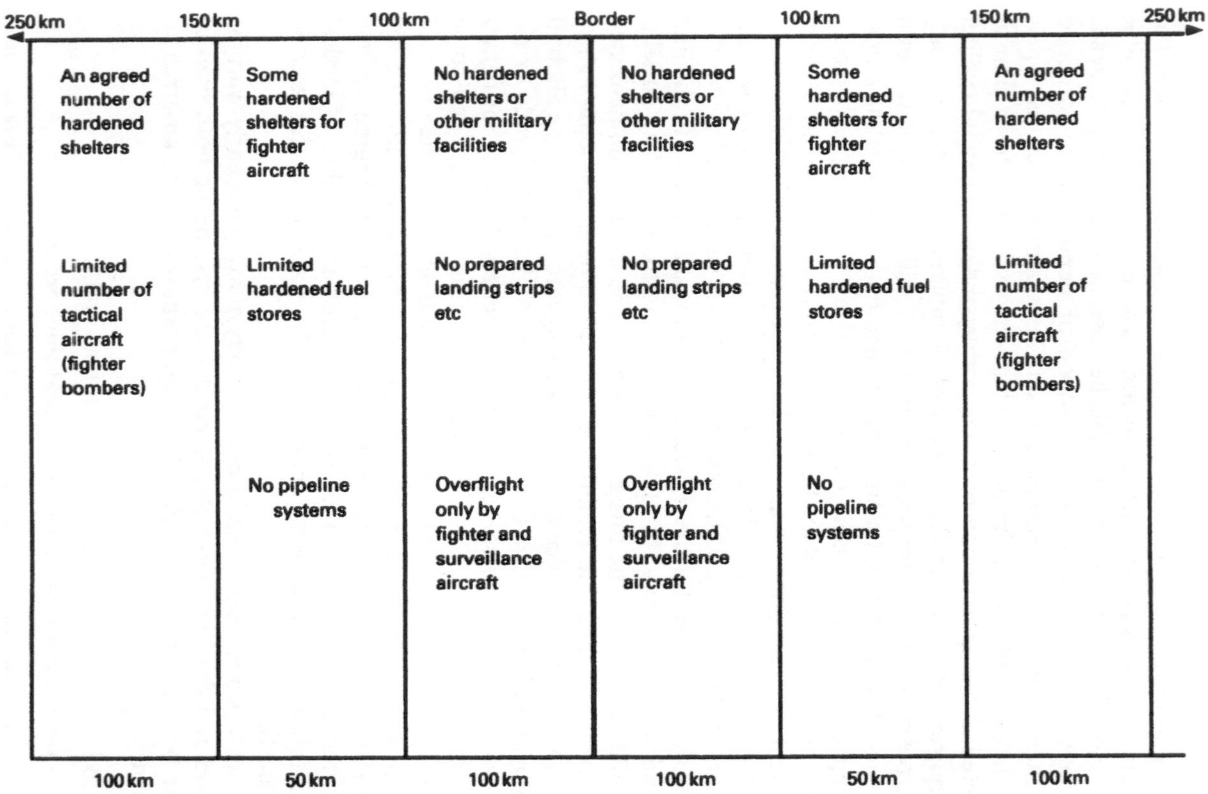

250 km / 100 km	150 km / 50 km	100 km / 100 km	Border	100 km / 100 km	100 km / 50 km	150 km / 100 km / 250 km
An agreed number of hardened shelters	Some hardened shelters for fighter aircraft	No hardened shelters or other military facilities		No hardened shelters or other military facilities	Some hardened shelters for fighter aircraft	An agreed number of hardened shelters
Limited number of tactical aircraft (fighter bombers)	Limited hardened fuel stores	No prepared landing strips etc		No prepared landing strips etc	Limited hardened fuel stores	Limited number of tactical aircraft (fighter bombers)
	No pipeline systems	Overflight only by fighter and surveillance aircraft		Overflight only by fighter and surveillance aircraft	No pipeline systems	

Figure 14.1 An 'open airfields zone'

- to enhance the effectiveness of defensive counter-air assets (such as surface-to-air missiles, anti-air artillery, fighters, and early-warning equipment),
- to establish an equal basis for further limitations on offensive air capabilities.

A system of zones as described is meant to function in conjunction with a zonal arrangement for ground forces with a first belt designed to wear down, delay and disrupt attacking units. In terms of overall strategic stability, the defensive arrangement for the ground forces is the main factor. It is also an important part of the prevention of air attack, for if air attacks cannot be followed up by ground forces their military value is in most cases relatively modest. Such attacks are in many cases more akin to acts of terrorism.

ASSOCIATED MEASURES

The main effect of an 'open airfields zone' is to improve early warning, a decisive factor for the effectiveness of air defence systems. With effective early warning, surprise attack on airfields is virtually impossible.

Improved early warning may become a necessity as foreseeable developments in aircraft technology materialise. Stealth technology, for example, can significantly reduce the detection range of radar – albeit at an enormous cost. The development of radar-evading aircraft could be an important incentive for establishing a relatively deep defensive zone with no important targets but with multiple sensor systems (radar, passive infrared and millimetre-wave radiometric sensors), even on a unilateral basis. Evidently there should be no restrictions on the deployment of early-warning systems of all kinds (ground-based and airborne) within the zone. Surface-to-air missile systems and anti-air artillery systems should also be allowed within the zone without limitation on numbers, provided they cannot reach enemy air space to any significant depth and interfere with airborne early warning. If early warning is adequate it is sufficient from a military point of view to be able to react against incoming aircraft as soon as they cross the border.

Effective surveillance of ground activities in the opposite zone is an important part of a zonal arrangement. Satellite systems and airborne surveillance across the border (with Synthetic Aperture Radar or other means) are one element of this. Prohibition of the use of electronic and other means of interfering with surveillance and early-warning activities should be part of a zonal arrangement.

But an 'open airfields zone' agreement would presumably include provisions for verification which amount to a form of on-site surveillance. In fact, there is no sharp distinction in this case between surveillance, verification and confidence-building measures. An 'open airfields zone' could be rather easily verified through permanent observers or *ad hoc* visits to relevant airfields within the zone. If there are also agreements limiting the deployment of offensive ground forces in certain zones it would be natural to supplement observation at airfields with an 'open skies' arrangement.

An 'open airfields zone' is designed to make airfields near the border ill-suited for military operations and vulnerable. This would encourage both sides to deploy their attack aircraft relatively far back, which facilitates early warning and air defence on the other side. It could also lead to accelerated development of alternative means such as Vertical/Short Take-off and Landing (VTOL/STOL) aircraft and long-range missile and rocket systems for offensive counter-air and other deep strike missions against fixed targets. A zonal arrangement may have to include limitations on such systems.

Concerning surface-to-surface missiles and rocket systems, the main consideration is how they would affect the balance of offence and defence for ground forces. Short-range surface-to-surface systems, for example missiles and multiple-launch rocket artillery with ranges up to 50 or 100 kilometres, would be an important part of a defensive set-up for ground forces, as they permit almost instantaneous concentration of fire to counter enemy concentrations. With improved munitions such systems will make a breakthrough with ordinary offensive ground forces extremely difficult.

From the point of view of the stability of an 'open airfields zone' the main point is to ensure that air defence assets do not become vulnerable to surprise strikes by long-range surface-to-surface systems (cruise and ballistic systems) or similar air launched systems. For this reason, if no other, such systems may therefore at least have to be limited as part of an 'open airfields' agreement.[1]

Note

1. This chapter is an integrated version of two papers by the authors submitted to the Pugwash Study Group on Conventional Forces in Europe at its Sixth Workshop, held at Altamura, Italy, from 1 to 4 October 1987.

15 Possible Ways to Stabilise the Balance in Tactical Aviation on the European Continent
Alexander Konovalov

Theoretically, the reduction of offensive capability in the combat potential of tactical aviation is more difficult than the establishment of non-offensive structures for the land forces. The difficulty has many explanations. Primarily, the inherent high mobility of aviation is a factor which makes it difficult to confine it to a given region. This difficulty, however, can be overcome if the deployment data of all units of land-based tactical aviation are available and inspection of air bases is permitted.

A frequent objection in the West to the inclusion of tactical aviation in the category of offensive weapons is that aircraft cannot seize and hold territory, which, it is claimed, is the normal goal of aggression. This is an obvious fact. It should be noted, however, that with the proper use of tactical aviation it is possible to seize and hold territory using substantially less tanks and motorised rifle units. Moreover, recent experience shows that the seizure of territory is not necessarily an immediate goal of aggression. There are situations where the destruction or severe damage of vitally important targets of the opposing side is the goal of an aggressor's political and military policy. These are the tactics which the United States used against Libya. In such cases, tactical aviation becomes the most effective means of aggression.

Tactical aviation can be divided into a defensive category (air defence interceptors) and an offensive category (fighter-bombers, attack aircraft) which due to their guidance equipment and armaments are suited for ground attack. This distinction is often difficult to make because (particularly in NATO) aircraft and their crews are equipped and trained to carry out both missions. In the Soviet Union and other Warsaw Pact states, the same tactical aircraft are used with modifications for interception and ground-attack missions.

At the same time, there is an important difference between NATO and the Warsaw Pact regarding tactical aviation. The readjustment of aircraft and the replacement or retraining of crews from interceptor to ground-attack missions in the Warsaw Pact states would be a long process requiring the utilisation of special facilities, taking at best several months. The same conversion by NATO can be done at the air base and requires a few hours (replacement of electronic guidance modules and hook-up of air-to-ground weapons); and, furthermore, no replacement of crew is required.

There also exists a historical difference in the philosophy of tactical aviation between NATO and the Warsaw Pact. The West traditionally prefers heavy multifunctional and more expensive machines, whereas the Soviet Union has for a long time preferred relatively simple, specialised types of aircraft.

All these differences demonstrate the difficulty involved. Yet they also highlight the fact that without the resolution of this problem it would be difficult, if at all possible, to speak of stable non-offensive military postures. Therefore, as a first step, let us examine some suggested methodologies. These could be used as a foundation for the development of a mutual understanding of the problem between NATO and the Warsaw Pact.

The most easily acceptable and presentable is the method of adopting common ceilings. To account for the quantitative and structural asymmetries in the tactical aviation of the two sides, it is logical, in our opinion, to use two parameters: the quantity of the aircraft; and the total combat payload, deliverable at a certain range. The second parameter is of vital importance for determining the realistic combat capability of tactical aviation and its role in the development of large-scale offensives. Logically, it corresponds to parameters such as the throw-weight of Intercontinental Ballistic Missiles or Submarine-launched Ballistic Missiles whose reduction by 50 per cent is envisaged in the negotiations for an agreement on a reduction of strategic offensive nuclear weapons. This suggested parameter for strategic missiles has disturbed the United States and it is understandable that the Soviet side feels the same with regard to this parameter in tactical aviation. The necessity of limiting the combat payload of tactical aviation is, incidentally, mentioned in the proposals developed by the American expert Ronald Hatchett.[1]

Of course, the payload of an aircraft is intimately tied to its operating range, and depends on its in-flight refuelling capability and flight path. For example, flying at low level with frequent manoeuvring through organised air defences increases fuel consumption and lowers the operating range of the aircraft. It is best to determine the operating range by a generally accepted formula which does not include the in-flight refuelling capability of the aircraft. The range is, in this case, determined by the optimal flight path under a given combat payload. In this case, a reduction in the offensive capabilities of tactical aviation could consist of a number of distinct phases.

First, there could be an agreement on what constitutes strike aircraft within tactical aviation, that is the aircraft with ground-attack capability. For example, the multipurpose aircraft that can be readjusted for a ground-attack mission by replacement of guidance systems and on-board weapons at their permanent bases could be categorised as strike aircraft. But those aircraft which perform interception missions and require extensive modification at special facilities in order to be readjusted for ground attack could be considered interceptors. In the first phase, the mutually acceptable limit for tactical strike aviation would be set at M aircraft.

In a second phase, the agreed number of ground-attack aircraft could establish a basis for determining the number of permitted interceptors. The actual calculations could be along the following lines: if the allowable number of land-based strike aircraft of each side does not exceed M, then the allowable number of interceptors for the opposite side should not exceed KM, where K is the coefficient of interceptors permitted for side A per one strike aircraft for side B. For example, if the agreed limit for strike aircraft, after reductions, is 1000 for each side and $K = 2$ then the allowable interceptor limit is 2000 ($2000 = 2 \times 1000$). If one of the sides reduced its strike aircraft to below the set limit, then the opposite side must also reduce its interceptors.

These two phases could be supplemented by agreements fixing the total combat payload of all 3000 tactical aircraft, that is both sides would be restricted to 3000 tactical aircraft with a combat load not to exceed L thousand tons deliverable within an operational range of 1000 kilometres. This criterion could be divided into two parts: total combat payload deliverable within a range of 500 kilometres and within a range of 1000 kilometres.

The above proposal for the reduction of tactical aviation and related structural changes would, in general, result in each side being allowed to deploy up to M strike aircraft, and no more than KM interceptors without ground-attack capability; and the total combat payload deliverable within an operational range of 500 kilometres would not exceed $L_{0.5}$ thousand tons, and in the case of those with an operational range of up to 1000 kilometres it would not exceed $L_{1.0}$ thousand tons.

These conditions could be supplemented by limitations on strategic interceptors, which would be determined on the basis of quantity of strategic bombers permitted to be deployed after the 50 per cent reductions in strategic nuclear weapons, that is

$$S_{si} = K_{sb}M_{sb}$$

where:

S_{si} = the number of strategic interceptors of one side;
K_{sb} = the agreed number of strategic interceptors per one strategic bomber;
M_{sb} = the number of strategic bombers left on the other side, after the reduction.

This proposal does not include the Soviet medium-range bombers based in Europe, Soviet land-based naval aviation aircraft and the carrier-based aircraft of the Western countries. Although the carrier-based aircraft do not officially fit into the mandate of the Vienna talks, they exist within the NATO–Warsaw Pact balance in Europe and they have a serious influence on the southern and northern flanks. It is relevant to note that the United States and its NATO allies have about 900 carrier-based aircraft, which do not include reconnaissance aircraft or aircraft equipped with electronic counter-measures.

The Soviet Union has 52 light, short-range aircraft with a combat payload, in our opinion, not exceeding 95 tons. On the other hand, the combat payload of the 900 aircraft of the NATO countries, deliverable within a range of 1000 kilometres, is approximately 3200 tons. The total combat payload of US and West European carrier-based aircraft and the medium-range bomber forces of the US Air Force and of France (100 FB-111s and Mirage-IVs), as we see it, is comparable to the total combat load of all Soviet medium-range bombers and all aircraft of land-based naval aviation (Tu-16, Tu-22, Tu-22-M).

Of course, this approach may produce an objection from the West; an objection related to the point that not all carrier aircraft should be included in the European equation because many aircraft are intended for defending the carrier group and not for attacking ground targets. Soviet naval aviation and medium-range bombers, however, are also not all based in Europe and ground-based naval aircraft with their guidance and navigation systems and on-board weapons cannot attack ground targets. Thus there is sufficient reason to postpone the problem of the Soviet's naval aviation and medium-range bomber force until a later phase in the process of the reduction and the restructuring of tactical aviation.

The quantitative restrictions on tactical aviation could be supplemented by strict on-site inspections of air bases and by placing a ban on the permanent deployment of strike aircraft at air bases near the NATO–Warsaw Pact border, primarily in Central Europe. Considering the exceptional cost and complexity of modern aircraft, it should be possible to agree not to destroy a percentage of the reduced tactical aircraft but to place them in storage where they would be kept under strict mutual control in a partially disassembled state (that is, with sealed equipment, removed engines, removed wings, and removed landing gear).[2]

Notes

1. Ronald Hatchett, 'Restructuring the Air Forces for Non-provocative Defence', in Marlies ter Borg and Wim A. Smit (eds), *Non-Provocative Defence as a Principle of Arms Reduction* (Amsterdam, 1989), pp. 177–88.
2. An earlier version of this chapter was published in Andrei Kokoshin, Alexander Konovalov, Valentin Larionov and Valeri Mazing, *Problems of Ensuring Stability With Radical Cuts in Armed Forces and Conventional Armaments in Europe* (Moscow, 1989).

Part V
Navies

16 The Vulnerability of Surface Forces to Modern Stand-off Weapons
Elmar Schmähling

INTRODUCTION

The employment of sophisticated microelectronics for data collection, transfer, and processing will revolutionise the appearance of war in the years to come. Indeed, completely new and promising forms of conventional warfare by means of today's technology could be available now if the way of thinking of the military was not too deeply rooted in the experience and concepts of previous wars. Today, the technology of tomorrow is faced with the operational thinking of yesterday.

Fortunately, these new technological possibilities and their application are compatible with the tide of military and political considerations which is moving us towards restructuring armed forces so as to increase political stability. Moreover, in the long run politicians and military planners will not be able to ignore the fact that the employment of this sophisticated technology will make increased defence capability possible at constant or reduced costs. All things considered, then, the prospect for a future restructuring of the armed forces has improved considerably.

I do not want to elaborate here in detail on the way that technological innovations dramatically improve the possibilities for target detection and engagement. That they do so can be taken as a fact. The implementation of the new technology in sensors and conventional weapon systems will ensure a virtual one hundred per cent success rate in the engagement of targets in future. The inability of targets to escape detection results from their conspicuous nature (in the sense of emitting active radiation), their sharp contrast with the detection background, their size, and their low capacity for rapid evasion. These conditions apply, in particular, to the large and expensive weapon systems which constitute a high proportion of the combat capability of all armies today.

Given the approximate equality of the two alliances regarding technological standards in the fields of detection, electronic warfare, propulsion and sensors, the 'large systems' on both sides will lose out to the smaller and cheaper missiles with their engagement success and cost effectiveness. This fundamental statement does not need to be qualified even when account is taken of the capacity of anti-missile defence, for this capacity may eventually be defeated by saturation. The prospects for successful saturation are best when there is a homogeneous detection background, that is to say at sea, in the atmosphere and in space.

Armament planners and weapon-operators ought to draw the following conclusion from this estimate of the situation: instead of concentrating combat capability on a continuously decreasing number of increasingly expensive weapon carriers, they should favour the introduction of as many hidden 'weapon carriers' or 'weapon platforms' as possible, located as remotely as possible from the opponents. And the introduction of self-propelled and long-range ammunition would be even better.

THE SHIP AS 'WEAPONS CARRIER'

It is often possible to explain existing systems and structures of armed forces only by their historical development. The experience gained in past armed conflicts and the conclusions drawn have a greater influence on present military doctrines and equipment, as well as on the planning of future weapon systems, than actual conditions or predictable future conditions.

The application of present military philosophies, which were possibly correct in the past, prevents real qualitative leaps in thinking about strategy and armament. The consequence is that armed forces' structures and operational principles are lagging behind the actual state of the art, despite the employment of modern technology.

This applies especially to naval forces whose character, even more than that of ground forces, is influenced by emotionally held fundamental convictions. This is no new problem. Sir Basil Liddell Hart in his memoirs described his experience with British admirals during the interwar period:

to most admirals the respective value of battleships and aircraft was not basically a technological issue, but more in the nature of a spiritual issue. They cherished the battle-fleet with a religious fervour, as an article of belief defying all scientific examination. The blindness of hard-headed sailors to realities that were obvious to a dispassionate observer is only explicable through understanding the place that the 'ships of the line' filled in their hearts. A battleship had long been to an admiral what a cathedral is to a bishop. [1]

Today, preconceived notions regarding the role of navies include the view that a state with a coast needs a navy; and that maritime areas can generally only be dominated and defended on sea. (I shall disregard the natural role of vessels for communication and trade across the seas.) Are these assumptions still correct? It has traditionally been the essential function of the warship, similar to that of the military ground vehicle or aircraft, to carry a weapon into the vicinity of the object to be engaged. Now, however, we have succeeded in improving the propulsion of ammunition to the point where it can be projected over large distances, to practically any location on the globe without an additional (manned) vehicle. This is technically feasible and economically acceptable. Another reason why one had to close in on the enemy target in the past was the short range of the sensors for target acquisition and discrimination. But already today, and even more so in the future, it is possible to locate and identify any target on the sea at any location and under all environmental conditions. It follows that the domination of a maritime area, that is its use and its denial to another power, does not necessarily require the presence of friendly naval forces in the area.

THE CHANGED THREAT

It is maintained by military planners and politicians of the North Atlantic Treaty Organisation that they must be in a position to fight a prolonged war with the Warsaw Pact states. This idea is the basis of the plan to deploy military assets and forces from the United States to Europe. To meet this contingency a high requirement of cargo capacity is assumed in order to supply the forces and population in Europe.

Again, resulting from the experience of the last war (the 'Battle for the Atlantic'), naval forces are to be employed to protect merchant and resupply shipping. Two considerations are relevant in this context. First, to destroy ships (transporters and escort ships) enemy surface units are no longer required on sea. It must be assumed that the future threat will emanate from the ground, from aircraft or from submerged submarines. Due to their long range and the use of indirect target acquisition, the launchers cannot be engaged by the escort ships. The latter's value can therefore derive only from their capacity to mount a successful defence against missiles. So the question arises whether the transporters themselves should be equipped with anti-missile systems.

Secondly, the assumption that there will be a prolonged war is the basis of existence for some NATO naval commanders. To assume a prolonged war, however, runs counter to the political aim of preventing war altogether. It is a declared principle that neither a nuclear nor a prolonged conventional war would be acceptable to West Germany. But consideration of currently conceivable scenarios of crisis development between the Warsaw Pact and NATO in Europe makes it seem likely that any overseas reinforcements would be too late for this objective. Hence, if the philosophy of war prevention (deterrence) is to be logically applied, the concept of reinforcing Europe will have to be given up in favour of an improved, fully manned and fully equipped defence capability. Any defence capability which is not already available in Europe in peacetime does not have a deterrent effect.

NAVAL FORCES AND CRISIS MANAGEMENT

Surface targets are particularly easy to discover and to engage. In a duel at sea what matters is who is *first* to locate and engage the enemy effectively. For this reason, the development of weapons and locating systems for naval assets has turned into a dramatic race for hundredths of a second. Marginal improvements have to be bought by exponential cost increases. The sensitivity of the equipment on modern warships means that it is no longer necessary to sink a ship when engaging it. It is enough to incapacitate highly sensitive electronic sensors and components. Consequently, surface ships will be in a 'high noon situation' in a future crisis. To continue waiting in a deepening crisis, while the opponents' weapons are trained at each other, is no longer acceptable from a military point of view. In a crisis, commanders in charge of valuable forces, such as carrier forces, must employ their weapons in response to threatening electromagnetic radiations, for these may precede an employment of weapons. For this reason, naval forces are much less suited for crisis management than the current doctrine implies.

This does not apply, however, to submarines. Since water still greatly complicates detection and target tracking, submarines cannot be eliminated pre-emptively, or only with great difficulty.[2] Hence surface ships, especially large and vulnerable ones, do not have a convincing role any longer in a war against an opponent who possesses a gapless reconnaissance system and a gapless weapons coverage. They are left in the absurd position where essentially they are driven to seek only self-protection.

This is not to imply that surface vessels have entirely lost their value. For, in accordance with their self-images, the great powers may wish to continue to use their naval forces to project power all over the world in low intensity conflicts. But in this context it is hard to see any justification for battleships.

It is not easy to foresee all the consequences for the future of navies. But so far as the defence of straits and coasts is concerned, the answer in certain conditions could be many small, possibly camouflaged and/or mobile weapon carriers on the coast or off-shore.

Effective defence of merchant ships against missiles and torpedoes will probably require a self-defence capacity. This might consist primarily of an anti-missile defence. But it could be supplemented by an anti-submarine defence in the form of helicopters stationed on merchant ships. For even in years to come modern torpedoes will not outrange anti-submarine helicopters (about twenty sea miles). From this it follows that the torpedo-employing submarine will have to move into the engagement range of the helicopter platform. For some time in the future, however, the torpedo will be of importance against merchant ships due to its high explosive charge and the mode of action/detonation under the ship.

To return to the broader theme, we may conclude that to increase crisis stability naval forces need to introduce weapon systems which

- themselves do not represent exposed and highly vulnerable targets;
- can destroy highly effective enemy systems (in case of an attack);
- are available in large numbers at at a low unit cost.[3]

Notes

1. Basil Liddell Hart, *Memoirs* vol. 1 (London, 1965), p. 326.
2. This advantage for submarines, even nuclear-propelled ones, may not prove to be permanent. For survival is becoming more difficult due to improved detecting devices and increasing possibilities for strategic anti-submarine warfare.
3. This chapter is based on two papers submitted to the Pugwash Study Group on Conventional Forces in Europe at its Fifth Workshop, held at Castiglioncello, Italy, from 9 to 12 October 1986.

17 Maritime Defence Without Naval Threat: The Case of the Baltic

Anders Boserup

Discussions of naval doctrine seem to be totally dominated by the classical notions of sea-power and destruction of commerce. According to these concepts, in any war the Soviet Union would seek to imperil the transatlantic supply lines, compelling the West to strengthen its escort capability, and, as in the strategy of John Lehman, former US Secretary of the Navy, to reassert undisputed naval supremacy, pushing the Soviet navy back through the Norwegian Sea and into closed seas like the Baltic, and tying it down in the defence of its home and base areas.

In the opinion of the present writer the concept of sea-power, of sustained control of the flow of goods over the globe, loses all meaning in a conflict in which the superpowers are in direct confrontation. Underlying the ideas of a second 'Battle for the Atlantic' and of striking the Soviet navy in its home bases, is the strange notion of a protracted conflict which is at once limited (non-nuclear) in its means and unlimited in its aims. Be that as it may, this concept of a naval contest between the superpowers in the Atlantic has dominated Western thinking for decades, and, with it, Western (certainly Danish) thinking in relation to the Baltic. The assumption that the Soviet Baltic fleet was destined for an out-of-area role suggested that control of the Danish Straits must be a major Soviet war aim and encouraged the notion of Western naval missions pushing deep into the Baltic in the event of a conflict.

The implication of these views is that the narrow sea-lanes at the Baltic Gate are of critical importance as the natural forward defence area for both sides and would be the focal point of naval, air and amphibious confrontation in this region in the event of war. Such notions seem to be out of touch with reality. For both sides it is relatively easy to close the Baltic Gate and extremely hard to keep it open. In the Danish area the waters are shallow and the sea-lanes long and narrow, and never more than a dozen kilometres off-coast. Even a moderately well-designed defence in these parts can make

179

transit by the Warsaw Pact navies virtually impossible, and it must be assumed that military planners in the East are aware of this. They cannot be planning for a blue-water strategy with units based in the inner Baltic. The Warsaw Pact, likewise, can pose a formidable threat to Western naval forces in the Baltic. If the West has absolute air supremacy the use of naval units for deep-strike operations in the inner Baltic is superfluous; if not, it is suicidal. In short: in case of a war in Europe neither side could hope to stage offensive action through the Danish Straits in circumstances when such action could still make a difference.

The main point of this chapter is to argue that the fact that both East and West can close the Baltic Gate is a desirable situation which it is in the interest of both sides to perpetuate and consolidate. Ideally, the Baltic Gate should have a double-lock so that either side can close it on its own, and so that it can only be reopened by common consent. Such a double-lock, based on highly survivable, non-threatening maritime defences on both sides, would have a decisive impact on security and military stability in this region:

- it would divorce the entire Baltic region from the destabilising and escalatory influence of the two-sided offensive naval postures in the Atlantic;
- it would stabilise the northern part of the Central Front by precluding seaborne outflanking operations on either side;
- it would make it pointless to commit large forces to the area and would thus divert military pressure away from it.

If the link to the Atlantic is severed, naval units in the Baltic would be reduced to supportive roles in relation to the adjacent land areas (landing operations, flank assault and logistic support). With effective and survivable maritime defences at or near the Gate the surface fleets would be of little use, even for those roles. Whether the Baltic Gate becomes a focal point of naval confrontation between East and West, or a key component in a stability-oriented, de-escalatory security regime in Europe, depends on whether defence postures can be designed which provide both sides simultaneously with a lock which could be expected to hold more or less indefinitely in a conflict.

To enhance crisis stability and make deliberate attack seem even less feasible than it does today, the main requirement is for defences which can be seen to be highly survivable and, in particular, invulnerable to pre-emptive attack. Near the potential front line the concept of trying to hold out for a short period while reinforcements are hastily brought in, does not offer credible deterrence, and in a crisis its consequences could be disastrous. The mainstay of a maritime defence in this area must therefore consist of predeployed forces which can be activated very quickly and which rely for survival on dispersion and stealth (perhaps also on hardening in a few cases). Surface fleets cannot be an important part of such a defence because they could not be relied on to survive beyond the first few hours against a massed sea/air attack. To try to lock the Gate by naval means is to invite pre-emption and to revive speculations about a possible connection between the Altantic and the Baltic. Both these would encourage a naval arms race and the adoption of offensive doctrines in the Baltic.

The same is true of aircraft. Equipped with modern stand-off missiles, they are a powerful means of sea interdiction in the narrow waters at the Gate itself and further into the Baltic, but both the aircraft and their bases are very vulnerable, and they certainly cannot survive for long on both sides simultaneously. Anyway, it is difficult to see how aircraft can be held back in a conflict, and, instead of taking the initiative, wait for an attack to develop, as they must if they are to provide a permanent defence. With airborne defences there may well be two locks to the Gate, but this does not help stability if the key that locks from one side can just as well be used to break open the lock on the other side. In fact, in this case as in others, the aircraft component of a defence array raises the spectre of pre-emptive attack and crisis-instability in an acute way, and the more so, the more exclusively the defence relies on aircraft.

The core of a stability-promoting maritime defence at the Baltic Gate must therefore consist of predeployed land-based forces and of rapidly deployable subsurface means: mines or submarines, depending on the circumstances.

Geographical conditions are particularly favourable for such a defence on the Western side, especially in Denmark. Moreover, the forces such a defence must be able to contain are of limited and known size. For the West maritime defence has to meet three requirements:

● it must make the straits impassable for a more or less indefinite period;

- it must make enemy shipping virtually impossible in the area west of Rügen;
- it must be able to prevent amphibious assault against the Danish Isles;

To meet these requirements a land-based (and subsurface) maritime defence system in the Danish area and in Schleswig-Holstein could consist of the following elements:

- Shore-launched, guided torpedoes. These would provide an affordable kill capability which can be disseminated in considerable numbers all along the coast.
- Sea-mines to obstruct and delay enemy movements.
- Dispersed, truck-mounted anti-ship missiles. These provide long-range kill capability (80–100 kilometres, say). If deployed in depth they combine stealth and survivability with instantaneous concentration of fire.
- Attack helicopters with air-launched anti-ship missiles. These provide the mobility and concentration needed to repel amphibious assault and a capability against sea-skimming vessels (air cushion landing craft).

Coastal artillery could perhaps also play a role. Its main defect is that it exposes itself when firing.

A dispersed and essentially land-based maritime defence of this kind would be very hard to suppress and would have no significant offensive capability. Vessels would be needed only for peacetime surveillance and, perhaps, for mine-laying. In this area the shipping lanes are so vulnerable to attack from the coast that the Warsaw Pact states would have to roll up practically the entire defence array in the southern parts before they could contemplate sea-borne transit or landing operations. If this defence array is combined with a reasonably effective defence of the airspace and with dispersed land forces to impede search-and-destroy operations, it should be possible to interdict the sea-space almost indefinitely.

Geography does not favour the Warsaw Pact states to quite the same extent. In their part of the Gate the seas are deeper and wider and they control only the southern shores. While dispersed, land-based maritime defences could cover the coastal waters effectively, they could not very well prevent warships from penetrating the Baltic. They would also be less survivable, at least in the area of the

German Democratic Republic (GDR), than their counterparts in the Danish Isles because they could be rolled up by forces coming in over land.

As if to compound the difficulties, the magnitude of the potential naval threat to the Warsaw Pact states is not well defined, and worst-case assumptions might therefore be expected to prevail. If the prospect of Western offensive naval operations deep into the Baltic is perceived as a significant threat, the Warsaw Pact has three ways of meeting it:

- with submarine forces operating in the area south of Sweden;
- with air forces, provided air superiority over the Baltic can be maintained;
- with amphibious assault forces which could seize points in the southern-eastern part of the Danish Isles from which to obstruct the seaways through the straits.

Of these it is the first which seems most practicable. It is also the least destabilising because it could not interfere with Western defences and could not itself be countered or suppressed by the West.

The areas which could be covered by the land-based maritime defences of the two alliances are shown on Figure 17.1. There is apparently a 'contested zone' between the southern Danish Isles and the GDR where the two areas overlap. In fact this overlap is a stabilising feature, similar in function to the fire-belt which Albrecht von Müller and others have been advocating for the Central Front because it encourages both sides to remain on the defensive. It should actually be regarded as the logical continuation of that belt on the northern flank. It makes all outflanking operations impossible and it should dispel all fears of the Baltic as a 'conveyor belt' for Warsaw Pact forces reaching right down to Kiel and Hamburg.

The map also shows where a Warsaw Pact submarine defence might be located. East of Bornholm the waters are eminently suited for hunter-killer submarines. Further west they are less ideal, and West of Rügen submarines can hardly operate at all. From the point of view of the Soviet Union, a submarine barrier in the Bornholm area is a forward defence. From NATO's point of view, a submarine presence in this area could not reasonably be construed as a significant threat.

Finally, the map shows two areas which could be militarily attractive for the Warsaw Pact, even in the context of strictly defensive

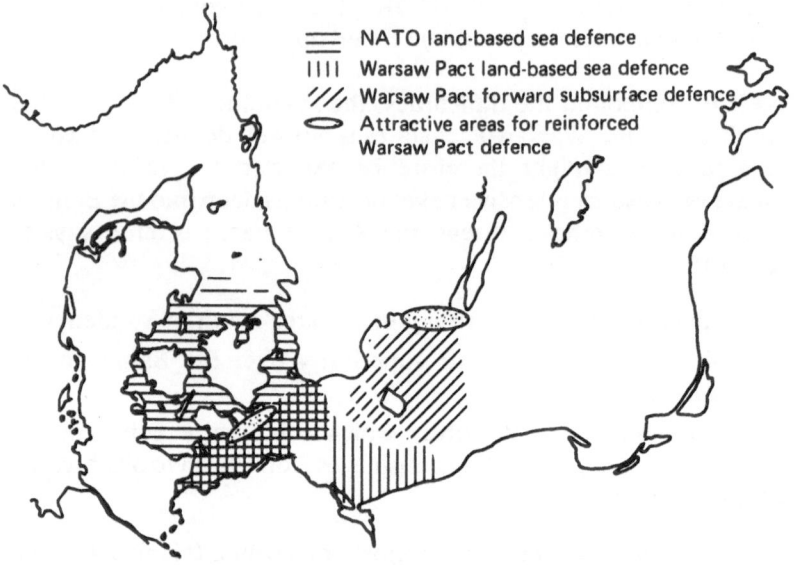

NATO land-based sea defence
Warsaw Pact land-based sea defence
Warsaw Pact forward subsurface defence
Attractive areas for reinforced
Warsaw Pact defence

Figure 17.1   Baltic maritime defences

aims in the Baltic region, namely, the rocky creeks of south-eastern
Sweden which are ideal as a forward base for submarines; and the
south-eastern Danish Isles, where control of a few coastal sites could
perhaps strengthen the Warsaw Pact lock on the Baltic Gate.

As this last observation shows, even a Warsaw Pact strategy which
is strictly defensive in the wider East–West context can contain
elements which are provocative, even threatening, in a regional
or local perspective. Or to put it differently: the US quest for
unchallengeable naval supremacy indirectly jeopardises the security
of its smaller front-line allies such as Denmark and of innocent
bystanders like Sweden by fostering threats which otherwise,
perhaps, would not have existed.[1]

Note

1. This chapter was submitted to the Pugwash Study Group on Conventional Forces in Europe at its Fifth Workshop, held at Castiglioncello,
   Italy, from 9 to 12 October 1986.

Part IV
Negotiation

18 *Glasnost*, Talks and Disarmament
Georgy Arbatov

Whereas in Soviet domestic affairs *glasnost* developed simultaneously with restructuring, and sometimes even ahead of it, in foreign policy matters were more complex. There were a number of reasons for this: first and foremost, the restricted nature of information. Of course, in domestic policy too information has been mercilessly concealed in the past. But the changes were more rapid in domestic than in foreign policy affairs, and the contrast has been greatest with respect to military affairs.

To some extent this is natural and even justified: military affairs have their own specific character making it impossible to open up everything to general inspection or to ensure total 'transparency'. This is because in the international arena we have to deal not only with friends, but also with foes, with whom a real struggle is in progress, a struggle whose rules do not allow all one's cards to be put on the table; and it is also because many foreign policy affairs affected and continue to affect the interests of third countries so that there is a need for tact in their public discussion (and sometimes temporary abstention from such discussion). But a part was also played in some areas – at times in many areas – by habit, by the force of inertia, and even by departmental self-interest.

None the less, the main natural laws of our time began to prevail – *glasnost* began to embrace and develop in the foreign policy and military spheres too, and not only in the form of initially timid articles by journalists and formulations by scientists. What was more important was, so to speak, 'criticism through action' – criticism of the mistakes made in the past through the revision of former negotiating positions, as well as bold practical actions in policy. In parallel, there began a creative discussion, in the spirit of the new political thinking, of major foreign policy problems in official speeches, as well as the principled formulation, at the nineteenth party conference, of the question of the shortcomings and mistakes that existed in this sphere.

Arising from this formulation of the questions, a scientific and practical conference was held at the Foreign Ministry in July 1988 in which Soviet military leaders, scientists, and journalists took part

187

along with diplomats. I think that this conference was an unprecedented event in the development of *glasnost* in foreign policy. It will be an important milestone in the awakening of foreign political thought and the development of debates on important foreign policy issues. One would like to hope that a similar event will take place at the Defence Ministry.

This conference reminded us all that in the spheres of foreign and military policy also the traditions, procedures, and customs that became established in the early 1930s can in no way be regarded as normal for socialism. We somehow forgot for a long time, and only remembered when restructuring began and *glasnost* entered our lives, that the October Revolution of 1917 also meant a fundamental change in public access to those formerly closed spheres of life. With its very first foreign policy acts socialism gave rise to a new, open diplomacy, which officially replaced secret diplomacy and was pledged to the open conduct of foreign policy before the eyes of its own people and others. This was no mere slogan: Lenin and his associates took this problem very seriously, proceeding from the assumption that socialism gains strength from the masses' consciousness, from the fact that the masses know everything, can form opinions about everything, and do everything consciously. Therefore, they believed, and, it soon became clear, believed quite rightly, that open diplomacy would help socialism stand firm, conquer, and secure a worthy place in international relations.

Of course, open diplomacy did not exclude confidential talks or secrecy in the course of preparing negotiating positions or carrying out diplomatic soundings. But foreign policy in general, and even the debates preceding any given major political act, were open to the broadest public. In this respect the young Soviet republic was truly a unique state, strikingly different from all other states. Remember the fierce debates that took place in full view of everyone about the Brest peace, the struggle against the supporters of the concept of 'revolutionary war', the debates about the lessons of the war with Poland and the preparations for the Genoa conference, the public disputes of the 1920s about military doctrine and the main avenues of military building? Not to mention the information about the numerical strength and composition of the armed forces that was published under Lenin and, perhaps through inertia, for some time after his death.

All this lasted for a long time and was perceived by the party and country as a normal phenomenon. Far from harming Soviet foreign policy, it probably did much to help increase its effectiveness; far from undermining the defence capability of the Land of the Soviets, it strengthened it, although the country was at that time very weak militarily in comparison with its enemies. This foreign policy behaviour strengthened the political positions of the world's first socialist state, prevented it from diverting its limited resources excessively towards military needs, and helped to prevent major errors in diplomacy and in military matters.

These chapters in the history of Soviet foreign and military policy should be restored to us today, if only so that we know what was regarded as possible and normal in this area during the Leninist times of socialist construction, and what has entered our lives 'at the devil's prompting', so to speak. The burdensome, essentially anti-social traditions based on the absence of *glasnost* that were forcibly imposed in the years of Stalinism, when it comes to foreign policy, were for some reason virtually unaffected by the twentieth party congress of 1956. And in the period of stagnation – although in foreign policy not all the years assigned to this period were entirely stagnant for *détente*, and the first talks and the first agreements on strategic arms limitation arrived in the early 1970s – these traditions of ultrasecrecy, or rather deeply meaningful silence, became fossilised. They became fossilised in all spheres, but particularly with respect to questions of defence and foreign policy. Whole generations of people grew up who never considered and sometimes had no inkling of the damage that this practice does to socialism and to the country's interests.

The absence of *glasnost* in the foreign policy and military spheres is a legacy we are renouncing. In this respect the scientific and practical conference at the Foreign Ministry can be seen as an invitation to a serious discussion – in the press too – of urgent problems. One of these I would like to discuss in more detail. In his speech at the conference, Foreign Minister E. A. Shevardnadze also spoke of the need to further develop the theory and practice of the very institution of diplomatic talks. In this connection he mentioned studies under way at the United States of America and Canada Institute of the Academy of Sciences in Moscow. In this context I would like to set forth, by way of debating points, a number of considerations reflecting both certain conclusions from these studies and my own personal opinion, first and foremost on the subject of arms limitation and disarmament talks.

For some time now many people have been voicing dissatisfaction with the progress and procedure for conducting these talks. There is criticism of the fruitlessness of many talks, their long, drawn out nature, and the paucity of results, as well as the fact that they by no means always help to stop or even to curb the arms race, sometimes merely creating the illusion that something is being done and thus placating the public and parliaments. There are undoubtedly grounds for this criticism. Of course, this state of affairs is engendered first and foremost by the actual policy and intentions of the sides (or at least one of the sides) conducting the talks. But as time goes on, there is a growing impression that the actual 'model' of talks is also in some ways incorrect, in some ways inappropriate to the difficult tasks set.

This model (with some modifications, of course) is borrowed from trade, from the sphere of commerce. The outline is simple here. First, one makes the highest possible demand, sets the highest possible price, and then bargains stubbornly. In the course of this bargaining the sides take some steps to meet one another and to make mutual concessions. Finally, it all ends with an agreement. This model, which has stood the test of time, is doubtless not without logic. Moreover, it reflects something in the general law of human and political behaviour. Everyone wants the maximum, but since both in life and in politics one has to be satisfied with the possible, after trying to get everything, one then moderates one's appetite and agrees to a 'reasonable sufficiency'.

But the experience of the 1970s and the 1980s gives rise to doubts as to the expediency of this model, especially in relation to arms control and disarmament talks. Why? First, because this model, by its very nature, presupposes prolonged bargaining and, inevitably, much wasted time. This condemns the diplomatic process to a very slow snail's pace on disarmament issues. But meanwhile, military hardware and military science are developing at an increasing pace. As a result, we are doomed to discuss not so much today's and tomorrow's problems, as the problems of yesterday.

Secondly, the 'goods' at talks on arms limitation and disarmament are not cloth, or grain, or machines, but security, truly important national interests. It is quite natural that in making its initial proposals each side tries to prove to the other that what it has proposed is absolutely just and necessary to safeguard its vital interest (otherwise no one will take one's position seriously at all). But when this has been repeated many times over, and not only by the negotiators but also by political leaders in public speeches, it becomes politically very difficult to deviate from one's first, initially stated positions. This is especially so since negotiating positions are, as a rule, worked out through an internal struggle and represent a compromise between different internal interests and departments. And there are always parties that actually have an interest in thwarting an accord on a particular issue and sometimes even in wrecking the very possibility of agreement. (An example is the military–industrial complex, which has recently considerably strengthened its position in the United States.) Because of the concessions and deviations from the initial position, it becomes easier for these forces to compromise the accord in the eyes of public opinion or parliament, which will later have to ratify the treaty. In other words, the present model of talks actually has programmed into it serious political difficulties in the drafting and ratification of agreements, because the proposed concessions can be interpreted by any demagogue as a 'betrayal' of national interests.

Thirdly, because of the length and complexity of the talks, the very institution of talks of this kind inevitably becomes bureaucratised. Every negotiating table becomes a kind of independent department with its roots deep in the bureaucratic structures of other departments. Sooner or later it develops a tendency to become totally divorced from its initial purpose, and begins to operate for its own sake, in accordance with Parkinson's well-known law.

Everyone who has carefully watched the progress of disarmament talks for nearly twenty years has seen various manifestations of these negative aspects of the existing model. What can be proposed instead? I believe that experience is already suggesting a more suitable model. This is a model whereby, after a more or less prolonged period of consultations at the level of experts and senior officials of the foreign ministries and other departments, an accord about the issues at the highest level or a very high level (that is, at the level of foreign ministers) is first reached. This is justified in view of the enormous political and defence significance of the problems tackled at disarmament talks.

At this very high level, detailed directives could be issued to those whose task it is to translate the accord into the language of legally formulated agreements, and, moreover, in a very short space of time. After drawing up such an agreement, the negotiators would disperse and the talks would close until they received directives on any new problems. That is very important. In the absence of an accord in principle at the highest levels, talks should not take place at all. When that accord has been reached, they should begin again and continue until the drafting of the next concrete agreement, and so forth.

In general, this method has already been tested to some extent. The first attempts were made during the talks leading to the Strategic Arms Limitation Talks Agreements (SALT I and SALT II) though not always consistently. The Soviet Union raised this matter far more resolutely at the Soviet–American meeting in Reykjavik. Unfortunately, because of the American side's position an accord was not then reached. But this new approach proved entirely successful during the drafting of the Intermediate-range Nuclear Forces (INF) Treaty.

Another important aspect of this problem concerns *glasnost* in disarmament talks. The Americans, by virtue of a certain political calculation, and we Soviets, by virtue of long habit, have made disarmament talks confidential. For certain aspects this is perhaps logical and will remain so. But not for every aspect. The experience of the 1970s and 1980s offers plenty of food for serious thought in this respect.

First, the American side has by no means always observed the gentleman's agreement on confidentiality. On the contrary, by means of carefully organised 'leaks' it has frequently misinformed the public and presented a picture of the talks and the sides' positions in a distorted light to suit itself.

Secondly, it has happened that disarmament talks conducted under cover of confidentiality have turned into the opposite. This happened, for instance, with the talks on reducing armed forces and armaments in Central Europe. The question remains: was it worth playing along with the Americans for such a long time – the talks began in 1973 – in order later to share with them the dubious glory of being participants in fruitless fifteen-year talks? But we Soviets played along with them – if only by virtue of the fact that we played this game according to the rules imposed by the Americans. This is the kind of situation that must be avoided.

Thirdly, even at disarmament talks, in distinguishing between the confidential and the non-confidential, it is clearly important to remember that we have entered an area of *glasnost* which must be extended not only to domestic affairs but also to foreign affairs, including military matters and disarmament. Why has all that we tell the American negotiators been kept secret, as a rule, even from the experts who are charged with explaining and defending the Soviet position? Why do delegations of the US Congress regularly attend talks, but not delegations of the Supreme Soviet? Why, lastly, does the Supreme Soviet Foreign Affairs Commission only 'become involved' in the problem when the time comes for ratification of the treaty, but has no opportunity even to express its opinion on the progress of talks or on many other important political issues? Would it perhaps be useful for the Supreme Soviet also to set up a commission on defence questions? It is certainly time we tackled these and other, similar questions within the framework of the planned political reform, democratisation and the expansion of *glasnost*.

These considerations are also relevant to the talks on conventional arms and armed forces in Europe which we are seeking to begin. Indeed, these talks certainly cannot be conducted according to the old model. First and foremost, this is because the subject of the talks will itself be incomparably more complex. The discussion will deal with all kinds of armaments and armed forces, which will have to be compared according to some scale of comparison that must be found. Moreover, geopolitical, geostrategic, and other factors will have to be taken into account here to an even greater degree than in the talks hitherto conducted. And last, but not least, the talks will be conducted not between two countries, but with the participation of more than twenty countries.

Some of the problems arising from this are obvious even now. For instance, the fact that the agreement cannot be a single, all-embracing agreement on the model of the INF Treaty or the SALT I and SALT II Agreements. It is more likely to be a question of some kind of collection of agreements following a certain order and, in addition to resolving some specific problem, advancing the whole front of the talks. Indeed, the Soviet proposals outline in general terms just such an approach and even formulate certain components and stages of agreement. Other issues still await study and debate.

In international relations and foreign policy, and even more in military and disarmament affairs, many questions urgently require analysis and debate. This means realistic analysis and debate based on the principles of the new thinking. This 'model' for disarmament talks is only one of these issues, probably not even the most important. Ultimately, it is naïve to expect that through these talks, which are conceived by the West not as a means of ending the arms race but as some kind of instrument for 'controlling' it, we will reach a nuclear-free, non-violent world. That can come only from a radically different national military and foreign policy on the part of a sufficient number of influential countries, purposeful efforts on their part, an all-around strengthening of the United Nations, and other major changes. Talks will, of course, be an important part of the process even then, consolidating and giving treaty form to a consensus reached by the course of progress. But this, too, merits discussion, along with many other problems.[1]

POSTSCRIPT

This article, published over a year ago, provoked a useful debate. And some progress can be recorded. At the Malta summit the leaders reached a broad understanding and instructed the negotiators to work out the details. But the negotiating systems on both sides are a brake on progress. They remain bureaucratic and elephantine, in the hands of layers of agencies.

I am increasingly convinced of the importance of national actions, compared with negotiation, if the present opportunity for lower military spending and increased security is to be seized. The Soviet Union, by its national actions, has made great progress in ending the Cold War and reassuring the West that it need not fear attack by us. But many of the military cling to old approaches, and the West is only beginning to respond to our actions.

I am convinced that the Soviet Union is ready to go far along the road to disarmament. I call upon the West to test my country's sincerity with national actions and proposals to push forward arms reductions.

January 1990

Note

1.   This chapter first appeared in *Pravda*, 17 October 1988.

19 The Vienna Force Reduction Talks: Moving Toward Deep Cuts
Jonathan Dean

INTRODUCTION

Against the background of forty years of East–West confrontation, with Western suspicions of the Soviet Union kept alive by events like the Soviet invasion of Czechoslovakia in 1968, of Afghanistan in 1979, and menacing military manoeuvres at the time of the 1981 declaration of martial law in Poland, it is not surprising that the leaders of NATO were confused and sceptical about the sudden shift of Soviet policy which came with the emergence in March 1985 of Mikhail Gorbachev as Soviet leader. Given the interest shown by their own public opinion, NATO leaders found it expedient to accept Gorbachev's proposal of April 1986 for a new negotiation on NATO–Warsaw Pact force reductions. Yet they remained sceptical as to President Gorbachev's motives. Many doubted Soviet willingness to make far-reaching force reductions. Others thought the Soviet leadership was motivated by plans of its own General Staff to replace quantity with quality in Soviet armed forces, to cut their size, to reorganise them and to equip them with advanced weapons, making them a greater threat than ever.

As Gorbachev moved to make far-reaching concessions to the NATO-approved position in the Intermediate-range Nuclear Forces (INF) talks, a new source of Western concern joined the others. NATO leaders, especially those of the two European nuclear powers, Great Britain and France, became apprehensive that, through highly effective interaction between Soviet policy and Western public opinion fuelled by a continuing flow of Soviet negotiating concessions, many or all US forces might leave Europe and with them US nuclear arms. Then, the two governments feared, they might be faced by strong public pressures both in their own countries and in other NATO states, especially West Germany, drastically to reduce or even eliminate their own nuclear armaments. In either outcome, purposeful reorganisation of Soviet forces or a massive build-down, Western Europe would be faced by a continuing or still greater Soviet threat.

The Western governments reached a conclusion logical for the circumstances: they decided to use the new negotiations on Conventional Forces in Europe (CFE), which convened in Vienna in early March 1989, to administer a stiff test of the genuineness of the Soviet Union's desire to cut back its armed forces. Accordingly, the NATO proposal at the new Vienna talks called on the members of the Warsaw Pact to eliminate their very large numerical superiorities in major ground force armaments by reducing their holdings of these armaments to a new equal level. Until the test results were in, NATO would in effect maintain its current force level by proposing only limited reductions of its own forces in the new talks. If its fears proved justified, NATO's military position would not have deteriorated.

NATO reached this position in the late autumn of 1988. Even as it did so and during the ensuing six months, the Soviet leadership undertook a series of actions – including Gorbachev's announcement in December 1988 of large unilateral Soviet cuts and the positions advanced by the Warsaw Pact as the new CFE talks got under way in Vienna – which met NATO's test, providing convincing evidence of Soviet willingness to make deep cuts in their own forces. President Bush's initiative at the NATO Summit of May 1989 appeared to demonstrate that his administration accepted this evidence as genuine and his views seemed shared by other NATO governments.

But although NATO moved towards the assessment that the Soviets' interest in cutting their forces was real, it had not drawn the long-term conclusions for its own policy concerning the possibility of deep cuts in its own conventional and nuclear forces which this assessment warranted. In this chapter, we will review what was accomplished in the remarkably active initial months of the CFE talks prior to the first summer recess in mid July 1989, and what remains to be done to place the CFE negotiating process firmly on track of continuing the build-down of the East–West military confrontation in Europe, both in its conventional and nuclear aspects.

STATUS AFTER INITIAL MOVES

In the first four months of the Vienna talks, there was rapid movement, first by the Warsaw Pact and then by NATO. The two alliances reached agreement on the force components to be reduced: tanks, artillery, armoured troop carriers, combat aircraft, armed helicopters, and active-duty ground and air force personnel. (By implication, there is also agreement that tactical range nuclear armaments should be reduced in a separate US–Soviet negotiation. However, NATO wanted this negotiation to take place at a later stage; the Warsaw Pact, immediately.) The alliances also agreed on a reduction method – a percentage cut from the current holdings of the numerically weaker side, NATO in most of these cases, with the Warsaw Pact coming down to the new NATO levels. This progress in resolving the main conceptual issues in the talks was startling. It placed the participants in the CFE talks after only four months of negotiation about where the superpowers were in the INF negotiations after four years of negotiation.

In addition to this conceptual agreement on general approaches to reductions, in the short period between 9 March, when the CFE talks began, and the end of May 1989, Warsaw Pact negotiators indicated willingness in principle to accept five main elements of NATO's highly demanding reduction proposals. They were:

● Reducing the heavy main battle tanks of both alliances to 20 000.

● Reducing armoured troop carriers to an equal level of 28 000 per alliance.

● Reducing artillery to a common ceiling of 16 500. There were definitional differences on artillery and, to a lesser extent, in the other two armaments, but they were soluble.

● Placing a sub-ceiling on Soviet holdings of remaining armaments of the type reduced at 60 per cent of the overall total for the Warsaw Pact.

● Placing a 'stationed forces' sub-ceiling on Soviet holdings of tanks, artillery and armoured troop carriers deployed in Eastern Europe.

In the early phase of the talks, the Soviet Union had not agreed to the specific stationed forces levels suggested by NATO; there was disagreement over 6000 stored NATO tanks, mainly US tanks. But Soviet agreement in principle to the idea of such limits meant that a

CFE agreement probably would contain a ceiling for Soviet forces stationed in Eastern Europe which would be an important obstacle to forward movement of Soviet forces to build up for attack or to reinforce political control – an obstacle which might be decisive in the light of already high Soviet reluctance to intervene in Eastern Europe.

Against the background of the unilateral reductions announced by Gorbachev in December 1988, these far-reaching Soviet moves at Vienna were unambiguous evidence of a Soviet desire to build down the East–West confrontation in Europe. In response to these Soviet concessions, NATO too made some useful moves in the form of Bush's proposals at NATO's fortieth anniversary meeting of the heads of government of member states held in Brussels on 29–30 May 1989. The Bush proposals were endorsed on that occasion and became NATO policy.

AIRCRAFT REDUCTIONS

President Bush's proposal to reduce combat aircraft and helicopters was the most important of these moves. Since the outset of CFE, the Soviets had insisted that combat aircraft be reduced as part of an overall agreement eliminating the Warsaw Pact's large superiorities in tanks, artillery and armoured troop carriers. It is likely that the Warsaw Pact would have made Western agreement to include aircraft a requirement for any CFE agreement. Moreover, given the agreed aim of the new Vienna talks to reduce the capacity of either side to attack, it was wholly logical to include in reductions aircraft and helicopters, which would be essential to the success of any offensive operation.

Bush's aircraft reduction proposal eliminated an estimated two years of debate within NATO about when to move on this subject and thus brought an early East–West agreement closer. His suggestion that the aircraft cut be set at 15 per cent generally corresponded to the dimensions of the first-stage cut proposed by the Warsaw Pact and did not appear to raise problems as regards the size of the cut. But there was wide East–West divergence over which aircraft would be reduced by both sides and over who had more of them. The Soviet Union counted ground-attack aircraft and fighter-bombers on both sides, but excluded nearly all other types of aircraft deployed in the Altantic-to-Urals area, including interceptor and training aircraft.

Using the data which it published in January 1989, the Warsaw Pact appeared to be claiming that NATO has about 4000 aircraft of the type it had proposed for reduction and the Pact about 2800. Thus, in a 15 per cent reduction from the current level of the weaker side, the Warsaw Pact would have reduced about 400 aircraft to a level of 2400 whereas NATO would have reduced 1600 aircraft. For its part, NATO counted in its data all fighter interceptors and training aircraft, as well as ground-attack aircraft, fighter-bombers and medium bombers, for a Warsaw Pact total of roughly 14 000 aircraft to NATO's 6600 (including 'combat-capable' trainers). Using this reduction base, after a 15 per cent cut in the NATO level to a common ceiling of 5700, NATO would have reduced about 900 aircraft and the Warsaw Pact about 8000.

The big differences were in NATO's inclusion of interceptor aircraft and training aircraft; the Warsaw Pact had a large numerical superiority in both types. NATO insisted that Warsaw Pact interceptors could be used for ground attack, to obtain air superiority over a battle area, and to escort fighter-bombers in attacks over NATO territories. NATO also argued that Warsaw Pact trainer aircraft were combat aircraft set aside for training purposes and capable of use as combat aircraft at any point. For years, however, the Soviet public and the Soviet military had heard about Western air superiority. Clearly, it would be difficult for both to accept that aircraft were yet another weapon system for outsized Soviet and Warsaw Pact reductions. Moreover, aircraft are far more conspicuous and costly incorporations of military power than tanks. Reaching agreement on reducing aircraft seemed likely to become the most difficult single negotiating issue of the CFE talks, and one which might have to be resolved at a high political level.

MANPOWER REDUCTIONS

Bush's proposals at the NATO Summit that US and Soviet manpower in Central Europe be reduced was valuable because it cut through NATO's previous reluctance to include military personnel in reductions. This reluctance was based on the difficulty of verification and on the negative data experience in the Mutual Balanced Force Reduction (MBFR) talks. The Bush proposal partially met the Warsaw Pact desire that negotiated reductions should cover military personnel as well as armaments. The proposal foresaw a 10 per cent reduction of about 30 000 US personnel deployed in Central Europe as compared with a reduction of 300 000 forward-deployed Soviet personnel down to an equal level of 275 000 men. The Soviets themselves proposed a stationed forces personnel limit of 375 000 in Central Europe. But they included in the Western quota more than 100 000 personnel of French, British, Canadian, Dutch and Belgian units stationed in West Germany.

Moreover, the Bush proposal did not cover Soviet military personnel deployed in Soviet territory west of the Urals. From the Western viewpoint, these personnel should be reduced or limited to give NATO some control over Soviet mobilisation potential and to preclude circumvention through increasing the personnel of permitted units. And, without negotiated manpower reductions for all NATO countries, it would be very difficult for NATO countries other than the United States to realise savings from force cuts. Some compromise between these two proposals would be needed.

Together with the verification issue, these two issues, air and manpower reductions, would be the most difficult to resolve in reaching a first agreement in Vienna. Yet it was probable that they too would be resolved.

DEEP CUTS

The events of the first months of CFE negotiations showed that a first agreement along the lines proposed by NATO was feasible. Consequently, NATO should expand its overall reduction objective and develop specific plans and proposals for the second phase of cuts to which both alliances have committed themselves in general terms.

For the past several years, the present writer has been arguing for an overall programme of moving to a lower equal plateau between the alliances based on a 50 per cent cut in NATO's current holdings of the offensive armaments which have been selected for reduction by agreement between the alliances: tanks, artillery, troop carriers, combat aircraft, combat helicopters – and, in the future, surface-to-surface missiles – as well as a 50 per cent cut from the present level of NATO active-duty ground and air force personnel. Other more defensive armaments would not be cut and their relative strength in the reduced forces of both alliances would increase. This programme would be carried out over a period of ten years and should result in cuts of about 30 per cent in the budgets of NATO countries for European defence.

There should be further deep cuts in holdings of armaments needed more for offence than defence – those armaments already agreed for reduction. NATO's phase one proposal for reduction of these arms was limited to about 10 per cent of its current strength. These reductions would leave both alliances at a high equal level. This outcome would be a great improvement over the present situation of Warsaw Pact superiority. But with both alliances still well equipped with armaments suited for attack, NATO's programme would fall far short of a maximum contribution to crisis stability. In addition to measures for early warning and to contain force activities and deployments, further cuts should be made in these offensive armaments in order to move towards defence dominance in both alliances. In practical terms, this would mean cutting another 40 per cent of NATO reducible armaments after a first phase, adding missiles and overall manpower to reductions, and bringing the Warsaw Pact down to these low levels. The Warsaw Pact has said it is ready to go down an additional 25 per cent in a second phase from the equal levels reached in phase one. It should go still further.

Looking at Central Europe, the overall reduction I propose would leave each alliance about 4000 tanks, 2000 artillery pieces and 5000 troop carriers. About half the Warsaw Pact's quota could be Soviet, enough equipment to provide for about ten light tank divisions or seven traditional strength tank divisions, as compared to the thirty divisions which have been deployed there up to the late 1980s. I suggest that, in addition to these cuts, a Restricted Military Area would be established in Central Europe from which all armaments of the type reduced and major ammunition storage sites would be prohibited, but where each alliance would be permitted to deploy infantry, combat engineers, anti-tank weapons, scatter and pipe mines, rapidly emplaceable obstacles and other counter-mobility measures.

Taken together, these measures would create a situation where defensive armaments and troop dispositions can bring to bear greater firepower than attacking forces. Successful attack by either alliance using foward-deployed forces in place would seem excluded. This would mean a better forward defence for NATO at half its present force level.

THE NUCLEAR ISSUE

A scheme for deep cuts does not constitute a complete long-term programme for NATO. This requires at least three more things. The first of these is a concept for a residual or minimum Western nuclear deterrent. Here I am assuming that, even after deep cuts in Soviet forces, majority opinion in Western European governments will still consider some minimum nuclear force necessary to counterbalance Soviet nuclear power with regard to Europe. It is urgent that NATO countries work on this subject now and that they discuss it directly with the Soviets at a high level, perhaps at a Bush–Gorbachev summit. Otherwise, controversy over this subject in the West, hitherto fomented by the Soviets, may create deadlock on force reductions, both conventional and nuclear, and make both alliances unstable. I have in mind a limited force on each side, consisting in the West of sea-launched cruise missiles, aerial bombs, air-launched stand-off missiles under 500 kilometres in range and of tactical range surface-to-surface missiles. A negotiated ceiling on surface-to-surface missiles should also cover conventional armed missiles whose large-scale deployment by both alliances would threaten stability.

Secondly, we have to tackle more deliberately the issue of how to maintain a healthy NATO as a long-term transitional institution providing cover and encouragement for a Western European defence grouping which should some day become fully autonomous. We must define specific ways of promoting that integration process.

Thirdly, we have to think of some East–West security structure for Europe. Here the most logical course appears to be deliberately to build up the rudimentary institutions of the Conference on Security and Cooperation in Europe (CSCE) into permanent institutions, like a NATO–Warsaw Pact centre for coordinating confidence-building measures, observation functions, and inspections. It could also function as a risk reduction centre to deal with low-level incidents. I am thinking less of Gorbachev's proposal for a new CSCE summit conference to decide on these issues than of a process of converting the many ongoing conferences of the CSCE process into enduring institutions covering the spectrum of international relations in Europe.

With more agreement within NATO on long-term aims of the reduction process and with more clarity between NATO and the Soviet Union on the nuclear issue, it may well be possible not only to conclude an early first agreement in the CFE talks but also to move beyond that to successful negotiation of a second stage of deep cuts on both sides.[1]

Note

1. An earlier version of this chapter was submitted to the 39th Pugwash
    Conference on Science and World Affairs held at Cambridge, Mas-
    sachusetts, from 24 to 27 July 1989.

20 Conventional Forces in Europe: Deep Cuts and Security
Hugh Beach

The negotiating positions of the two alliances and the prospects for the CFE negotiations are analysed by Jonathan Dean in Chapter 19. The purpose of this chapter is to look beyond Stage One and consider what should follow.

Both sides in their opening positions recognise the need for further stages in which forces would be cut and restructured so as to reduce offensive and increase defensive capabilities; and support for deep cuts has been voiced by people who speak with authority, for example Les Aspin, the Chairman of the US House Armed Services Committee, and General Andrew Goodpaster, a former NATO Supreme Allied Commander Europe (SACEUR).[1] But despite this support for deep cuts, despite the prospect of rapid progress being made in the Stage One negotiations and the high stakes riding on the outcome, no systematic study appears to have been made either in the West or the East of the scale of reductions which might be appropriate nor how the reductions might be made.

MILITARY OBJECTIONS TO DEEP CUTS

The NATO opening position in CFE is based on the military judgement that NATO could not make substantial reductions without fundamentally altering its defence posture. The crucial stumbling block, which up to now has prevented any systematic consideration of deep conventional cuts in Europe, is the issue of Forward Defence (which remains politically indispensable) and the corresponding need to maintain an adequate density of forces along the West German border with the German Democratic Republic and Czechoslovkia. These considerations lead NATO to feel that there are certain force levels below which it cannot safely reduce. It has twenty-two divisions covering a frontage of more than 1000 kilometres. This means that Allied Command Europe already has insufficient forces to cater for adequate mobile reserves. Reductions could not cut very deeply before considerations of terrain and force-to-space ratios would become a dominant factor.

Part of the difficulty arises from the problem of coming to terms with the breathtaking novelty of the context. The starting point (that is, the position reached at the end of Stage One) will be a condition of *parity*, properly monitored and supervised. Even before Stage One begins (that is, as a result of the Soviet unilateral process), a measure of restructuring will have taken place on their side. For example, the Soviets say that, in addition to withdrawing six tank divisions from Eastern Europe during 1989–90, they will cut the number of tanks in all the remaining divisions: in tank divisions by 20 per cent (from 320 to 260) and in motor rifle divisions by 40 per cent (from 270 to 160); limiting to what is strictly necessary the number of attack weapon systems and changing their deployment to make them purely defensive. The Soviets say that in future their operational concept for a war in the West, instead of relying upon an early and massive counter-offensive into the depths of West Germany, will involve standing on the defensive for the first three or four weeks. The importance of a change of this kind, if fully implemented, is difficult to exaggerate. First, this is because NATO's concept for the Central Front, as is discussed in more detail below, has always been one of fighting to rearward within West Germany. But it is difficult to see how hostilities could ever develop if both sides stand initially on the defensive. Secondly, such a change, set in the context both of quantitative checking of force reductions, and of a cooperative security system of the kind likely to emerge from the parallel Vienna talks on Confidence and Security-building Measures, would virtually be self-verifying. Operational concepts of this kind find expression not only (or even primarily) in secret deployment plans, but much more revealingly in peacetime deployments, logistic preparations (forward dumping of ammunition, assault bridging, and pipeline stocks), training manuals, pamphlets and magazine articles, and the general run of exercises. The transparency measures proposed by the Soviet side in Vienna include regular exchange of information on the numbers, structure and deployment of all forces down to regimental level. Other measures would allow the discussion of operational policy; the exchange of official visits; lower thresholds for notification and observation of ground force activity; and measures for improving observation including aerial survey.

So by the time a Stage One agreement has come into effect, not only will the numerical correlation of forces have been transformed, but even more far-reaching changes will have overtaken the operational context. Almost all the factors which combine today in occasioning the feeling that there is an uncomfortable level of risk will have been changed out of recognition. Seen in this new light, the requirements both of forward defence and of force-to-space ratios may assume a rather different complexion.

FORWARD DEFENCE AND THE PROBLEM OF THRESHOLDS

The political directive to fight the battle as far to the East as possible is not to be taken, as in the past it sometimes has been, to mean trying to defend statically and right up to the inner German border. The key to resolving this apparent contradiction lies in the concept of the Covering Force, a normal and orthodox military arrangement. The Covering Force consists of mobile units deployed, very early in any period of alert, into the area between the inner German border and the forward edge of the main battle zone. This area varies in depth: in the British Corps area from 6 to 60 kilometres. The task of these units in the first instance is to cover the deployment of the main body. Thereafter their task is to observe, harass, delay and call down fire upon the advancing enemy for as long as it is militarily sound to do so. They are then withdrawn through the main defensive position, regrouped, refurbished and thereafter form an important element of mobile reserve.

The main defensive battle is to be fought by each national corps, within its own allotted boundaries, according to the terrain and to its own tactical concepts. It was the outstanding achievement of General Sir Nigel Bagnall to have secured agreement, at least on paper within the Northern Army group, to a unified concept placing much more emphasis upon mobility than upon any rigid defensive system. It capitalises upon the capacity for manoeuvre inherent in the highly mechanised forces now provided throughout NATO. It looks to the air to keep the enemy off the army's back; to support the land battle directly when presented with specially appropriate and lucrative targets; and to impose the greatest possible delay on selected Soviet follow-up forces. The objective is to allow concentration of forces at critical points with the aim of forcing local tactical successes, seizing the initiative, capitalising on mistakes by the other side and, if all goes well, to provide the wherewithal for a locally decisive counterstroke involving reserves of at least one or more armoured divisions acting in concert. It is a bold and attractive vision, doubtless long overdue. It is also a very far cry from any narrow understanding of forward defence read in a literal sense. Using the British example once again, the corps area from the forward edge of the main battle zone back to the rear boundary is about 140 kilometres in depth. Arrangements are made for the conduct of operations throughout the full depth of this area; that is, to a depth of some 200 kilometres from the inner German border. The army group commander establishes a coordination line behind which national corps may not withdraw without his authority. In no other sense is operational success denominated in spatial terms.

A more usable criterion of military sufficiency, at least where public discussion is concerned, relates to the *time* for which effective operations could be sustained. The British Statement on Defence Estimates for 1970 defined the concept of flexible response in these words:

> NATO strategy depends critically on maintaining conventional forces in Western Europe at a level which will give NATO an alternative to nuclear response against anything but a major deliberate attack; and which, if an attack on this scale should occur, would *allow time* for negotiations to end the conflict and for consultations among the allies about the initial use of nuclear weapons if negotiations should fail (my emphasis).[2]

What has never been defined is how much time is needed or is available, and opinions have varied. The same Defence White Paper, for example, concluded comfortingly: 'At present the level of these conventional forces is just sufficient for this purpose.' In 1985, the Defence Planning Committee of NATO in ministerial session in Brussels commented: 'The current disparity between NATO's conventional forces and those of the Warsaw Pact risks an undue reliance on nuclear weapons. This would be an unacceptable situation which we are determined to avoid.' In 1988, however, General Sir Martin Farndale, then Commander in Chief Northern Army Group, struck a pessimistic note: 'We are only talking about a few days, then NATO will have time to get her act together and agree to a nuclear fire plan ... You know a very great deal has got to go right for us today if we are to stop that initial breakthrough.'[3] General Bernard Rogers, then SACEUR, in his address to the Annual Conference of the International

Institute for Strategic Studies held in Berlin, in 1985, pinpointed the problem as an inability to *sustain* NATO forces adequately with trained manpower, ammunition and war reserve material. He summarised the requirement as follows: 'Credible deterrence requires NATO to attain a conventional capability that would give a *reasonable prospect* of frustrating a non-nuclear attack by conventional means.'[4]

From this discussion three consequences follow. First, the nuclear threshold, to the extent that it has ever been defined, is denominated in terms of time, not space. Secondly, there is room for a wide variety of views, even among official spokesmen, as to how much time is currently available, and how much is required. But if it were the aim, under a regime of deep conventional cuts, to do no better than is now officially seen as attainable, then that is an undemanding target. Thirdly, however, it would make much more sense, rather than to talk in terms of space or time, to adopt some such formulation as that of Rogers. The 'reasonable prospect' of frustrating a non-nuclear attack goes beyond the minimal requirement of preventing any quick easy gain of *fait accompli*, but stops well short of guaranteeing a successful defence, still less of prevailing.

THE RATIO OF FORCE TO SPACE

In the conduct of a successful defence the key is *cohesion*. Self-evidently this implies the ability to communicate effectively across the battlefield, and to maintain a unified design for battle; hence the crucial importance of Command, Control, Communiciations and Intelligence (C^3I). Hence also the importance of preventing disruption: for example, by airborne attack against rear areas or by major breakthrough. But the first principle, as explained by Farndale, is that of *mutual* support along the whole front 'to provide cohesion and to defend key terrain using the minimum forces possible'. In traditional terms mutual support implies interlocking arcs of fire employing direct fire weapons. The limiting factor has not for many years been the range of these weapons but of *intervisibility*. For instance, it is assessed that in the inner German border region 55 per cent of the terrain has sighting ranges of 500 metres or less, 28 per cent from 500 to 1500 metres and only 17 per cent over 1500 metres. Until recently, this factor was reinforced by the high proportion of the year when there is some obscuration by mist (not to mention darkness), the fog of war and the lavish use by the Soviet army of tactical smokes; but infra-red imaging equipment can now largely discount these. The rise and fall of ground and the existence of buildings and trees remain the chief determinants of interlocking fires, and hence of the density of forces needed for cohesion.

In assessing the prospects for deep conventional cuts the crucial question can thus be put as follows. In a situation where, for argument's sake, the number of NATO divisions has been reduced from twenty-two to fifteen (and the number of Warsaw Pact divisions in Eastern Europe is also fifteen), how under the rubric of a 'reasonable prospect' of frustrating conventional attack can the cohesion of defence – above all the principle of *mutual support* – be sustained in a way that the professional military will find acceptable? It might be a good idea to start by asking them. Until such time as answers are forthcoming, the following considerations may be profitable.

ALTERNATIVE CONCEPTS

Up to now forward thinking in the West has been dominated by two quite different philosophies. The first has been inspired by American concerns to harness technological advances to a more offensive stance. These have found expression doctrinally in the form of concepts for the land–air battle which emphasise mobility, counter-offensive capability and means for carrying the battle into enemy territory. The counterpart, in terms of weapon development, is the NATO concept of Follow-on Forces Attack (FOFA) which looks to the provision of systems for deep strike against Warsaw Pact forces, employing advanced conventional munitions delivered by artillery, rocket, missile or (for the most part) aircraft. Many, if not all, of the technologies involved in these developments, driven in large measure by the need to increase the effectiveness of conventional munitions and thus further to downgrade the salience of the nuclear components in flexible response, will have application to the new circumstances. Technological development has a momentum of its own and, almost by definition, eludes the net of arms control save in specific applications. But the more expensive, technically demanding and (by reason of their hair-trigger character) destabilising components of deep interdiction will be out of key with the general context here depicted.

The opposite school of thought is that inhering in such concepts as alternative defence, non-offensive defence, non-provocative defence or *Strukturelle Angriffsunfähigheit*. These are rooted in the belief that what is needed is radical unilateral change in the nature of NATO forces, with the emphasis placed on defensiveness, to break the cycle of competitive force improvements. It is contended that efforts to match like with like, between forces both of which are configured on the classic *Blitzkrieg* model, is a recipe for a continuing arms race and instability in a crisis. A manifestly defensive system unsuited to attack would not be perceived as a threat, would reduce the risk of conflict and would provide more effective resistance if attack should come. With the wish thus father to the thought, alternative defence approaches have been proposed along three main lines.

- Organised non-violent defence or social resistance, accompanied by total or partial elimination of existing NATO forces;
- Replacement of allied armoured formations by forces organised in a radically different way;
- Modification of present *Bundeswehr* and other allied organisations, retaining a traditional armoured component (albeit much smaller) but adding a forward shield of high technology defences.

In most such proposals it is urged that NATO should forego large and vulnerable force concentrations in favour of small, dispersed, concealed, force groups. Forces should be limited in mobility and range and not be capable of deep counter-attack (least of all on enemy territory). The aim is to dissuade attack through more effective denial of one's own territory rather than deterrence through threat of retaliation. It is noticeable that political considerations have been paramount; the belief that Soviet attack on Western Europe for purposes of conquest is of negligible likelihood; and the desire to eliminate, or at least radically reform, the present NATO strategy of flexible response with its reliance on first use of nuclear weapons if conventional defences fail.

Partly because of this element of wishful thinking built into the foundations of alternative defence, these concepts have not stood up well to technical analysis. There is too much reliance on deliberate immobility of defence forces, an obvious handicap where resources are insufficient. There is too much emphasis on technical fixes such as instant anti-tank ditches and pre-emplaced sensor fields. There is the question of expense; given insufficient resources to field at an adequate level the armoured forces on which they rely, planners do not want to subtract from these to invest in limited mobility forces of any kind. There is far too little recognition of proven military principles, such as the necessity for counter-attack. NATO military commanders understandably have not been prepared unilaterally to give up the obvious means of dealing with a Soviet breakthrough, pushing an invading force back or interdicting the attackers in their own bases. As Hylke Tromp has pointed out in an elegant critique of non-offensive defence, 'asking the military to minimise their mobility and to limit their range comes close to asking them to perform an unnatural act ... that is not what they see as their vocation'.[5]

These two approaches, land–air battle and alternative defence, have several things in common. First, they both emphasise the need for new force structures: more of the same or, in the new context, less of the same is most unlikely to be the best answer. Secondly, they both emphasise the importance of using technological innovation in creative ways, focusing as it happens upon the identical cluster of emerging technologies albeit in very different applications: remote sensing, survivable communications, target acquisition, discrimination and homing at submunition level, advanced conventional warheads and novel delivery means such as Search and Destroy Armour (SADARM) and stochastic mines. Thirdly, they have both been developed, and represented as efficacious, within the *existing* strategic context and correlation of forces.

THE SCOPE FOR NEW APPROACHES

Now that it is the expressed intention of both East and West to move, in a cooperative and recriprocal manner, towards postures which might be described as 'defence only sufficiency', these doctrinal approaches need to be looked at again. From the high mobility viewpoint it will be contended that, in order to cover the front and carry out the defensive mission, Allied Command Europe would be forced to conduct more mobile operations, giving ground in order to gain time and to discover the main lines of enemy attack, while holding a strong force in mobile reserve to contain and destroy them. In this model surveillance and target acquisition from the air (including unmanned vehicles), air attack of armour (including helicopters) and high mobility forces (including heliborne) will be of high salience. Richard Simpkins, in his seminal book *Race to the Swift*, regards this as an inevitable development in any case: 'The evolution of rotary wing forces is but one of at least five trends ... the first and most tangible is the way the combination of surveillance and firepower is putting paid to concepts based on high troop densities.[6] In short, this brings in the vertical dimension to resolve the dilemma of force-to-space ratios. Incidentally, this line of thought places a question mark against current proposals for including combat helicopters among forces for immediate reduction in CFE.

From the alternative defence viewpoint it will be held that, while non-offensive schemes perhaps did not carry conviction when proposed for unilateral application against the undiminished Warsaw Pact order of battle (despite much special pleading from factions of the net assessment industry), the new context provides exactly the right setting for increased emphasis on obstacle planning, ground-based sensors, territorially based defences and all the other paraphernalia of classical non-provocative theory. At the very least this thesis deserves re-examination. A useful starting pont might be the overall concept put forward by Andreas von Bülow with the help of a team of West German analysts.[7] This draws in turn upon ideas put forward in 1970 by E. A. Burgess, later Deputy SACEUR.[8] In the first version of von Bülow's scheme, applicable prior to any substantial force reductions, the basic idea is to separate out defensive forces into shield and sword. The shield, which operates in an area about 60 kilometres deep across the whole inner German border, consists of some 200 home defence regiments, all of them provided by the *Bundeswehr*. They would be organised as light infantry but integrated operationally with the allied corps structure. This would enable the light infantry units to receive formidable support from artillery, including Multiple-launcher Rocket Systems (MLRSs), armoured engineers, anti-tank guns, drones and guided missiles, anti-aircraft missiles and airborne troops. They would be formed from local conscripts and reservists. They would rely upon terrain which is poorly suited or completely unsuited to massed armour operations. They would close the gaps on either side of the terrain in which the sword forces would join battle. The sword forces, consisting of West German and allied armoured brigades, 'the old still untouched monostructure of NATO' are reserved for reinforcing the defence and especially for counter-attack. They do not have to line up with their high-investment weapons near the border. This gives them a freedom of action which they have never before possessed. All this has been worked up into a detailed force posture for the *Bundeswehr*, involving a draft commitment of twelve months active duty followed by fourteen years on the reserve. Additional equipment costs for the *Bundeswehr* would be small, and personnel costs unaltered. There would be no necessary changes in the force structures of the other members. Shield forces would be subordinated in war to the allied corps within whose boundaries they operated. The scheme is thought out in considerable detail and has great attractions. But more importantly, in a second phase, von Bülow's structural and tactical ideas have been reworked in the context of a regime of deep cuts, even more far-reaching than the example mentioned earlier in this chapter, and applied in a scheme applicable simultaneously in East and West. In addition to separating out the shield and sword, this plan would cater for three zones of disengagement of which the narrowest, say 60 kilometres on either side of the border, would be without main battle tanks, armoured fighting vehicles and armoured artillery. There would be no restriction on 'barrier type forces', light infantry, engineers with minelaying capabilities, anti-tank and anti-aircraft systems unsuitable for attack. FOFA, and indeed the whole concept of offensive counter air, would gradually wither away.

Von Bülow's scheme is a visionary one, naïve in parts and needing much refinement. Nevertheless, it has great merits. It makes use of old ideas whose time may now have come. It represents a fusion of the concepts of high mobility and of alternative defence. The concept of the manpower-intensive shield addresses directly the problems of forward defence and the ratio of force to space. In its second form it is a symmetrical scheme as between East and West, in a version more fully worked out than has been done before. As an instance of the sort of work which now needs to be done, it provides a shining example. It is up to the other nations, and NATO as a whole, to set on foot comparable research; and the matter is now urgent.[9]

Notes

1. Les Aspin, 'Meeting Gorbachev's Challenge', unpublished paper, December 1988; and Andrew J. Goodpaster, *Gorbachev and the Future of East–West Security: A Response for the Mid-term*, occasional Paper of the Atlantic Council of the United States (Washington DC, 1989).
2. British Ministry of Defence, *Statement on Defence Estimates, 1970*, Cmnd. 4290 (London, 1970), pp. 5–6.

3. Sir Martin Farndale, 'The Conduct of Operations in the Central Region following the Withdrawal of the INF', paper presented to the European Army Conference, Paris and London, 25–27 October 1988.

4. Bernard Rogers, 'NATO's Strategy: An Undervalued Currency', in International Institute for Strategic Studies, *Power and Policy: Doctrine, the Alliance and Arms Control: Part I*, Adelphi Paper no. 205, London, 1986, p. 6.

5. Hylke Tromp, 'Non-offensive Defence, Conventional Stability, and the Military Balance', paper submitted to the Seventeenth Pugwash Workshop on Nuclear Forces, held at Geneva from 3 to 4 June 1989, p. 6.

6. Richard E. Simpkins, *Race to the Swift* (London, 1985), p. 131.

7. Andreas von Bülow, 'Conventional Stability NATO–WTO: An Overall Concept', Hearings before US House of Representatives Armed Services Committee, Washington DC, 7 October 1988.

8. D. M. Pontifex and E. A. Burgess, 'A New Concept of Land Operations in Europe', *The British Army Review*, August 1970, pp. 13–20.

9. This chapter was submitted to the Pugwash Study Group on Conventional Forces in Europe at the Eighth Workshop, held at Porto Vecchio, Corsica, France, from 29 September to 3 October 1989.

Index